PERGAMON INTERNATIONAL LIBRARY
of Science, Technology, Engineering and Social Studies
The 1000-volume original paperback library in aid of education,
industrial training and the enjoyment of leisure
Publisher: Robert Maxwell, M.C.

Introduction to Geological Maps and Structures

THE PERGAMON TEXTBOOK
INSPECTION COPY SERVICE

An inspection copy of any book published in the Pergamon International Library will gladly be sent to academic staff without obligation for their consideration for course adoption or recommendation. Copies may be retained for a period of 60 days from receipt and returned if not suitable. When a particular title is adopted or recommended for adoption for class use and the recommendation results in a sale of 12 or more copies, the inspection copy may be retained with our compliments. The Publishers will be pleased to receive suggestions for revised editions and new titles to be published in this important International Library.

Other Related Pergamon Titles of Interest

Books

ALLUM:
Photogeology and Regional Mapping

ANDERSON & OWEN:
The Structure of the British Isles, 2nd edition

ANDERSON:
The Structure of Western Europe

BARBER & WIRYOSUJONO:
The Geology and Tectonics of Eastern Indonesia*

CONDIE:
Plate Tectonics and Crustal Evolution, 2nd edition

OWEN:
The Geological Evolution of the British Isles

PRICE:
Fault and Joint Development in Brittle and Semi-Brittle Rock

SIMPSON:
Geological Maps

Journals

Journal of African Earth Sciences

Journal of Structural Geology

Introduction to Geological Maps and Structures

JOHN L. ROBERTS, B.Sc., Ph.D., F.G.S.
University of Newcastle Upon Tyne

PERGAMON PRESS

Oxford · New York · Toronto · Sydney · Paris · Frankfurt

U.K.	Pergamon Press Ltd., Headington Hill Hall, Oxford OX3 0BW, England
U.S.A.	Pergamon Press Inc., Maxwell House, Fairview Park, Elmsford, New York 10523, U.S.A.
CANADA	Pergamon Press Canada Ltd., Suite 104, 150 Consumers Rd., Willowdale, Ontario M2J 1P9, Canada
AUSTRALIA	Pergamon Press (Aust.) Pty. Ltd., P.O. Box 544, Potts Point, N.S.W. 2011, Australia
FRANCE	Pergamon Press SARL, 24 rue des Ecoles, 75240 Paris, Cedex 05, France
FEDERAL REPUBLIC OF GERMANY	Pergamon Press GmbH, 6242 Kronberg-Taunus, Hammerweg 6, Federal Republic of Germany

Copyright © 1982 Pergamon Press Ltd.

First edition 1982

Library of Congress Cataloging in Publication Data
Roberts, John Léonard 1936-
Introduction to geological maps and
structures.
(Pergamon international library of science,
technology, engineering, and social studies)
Bibliography: p.
Includes index.
1. Geology—Maps. 2. Geology, Structural.
I. Title. II. Series.
QE36.R63   1981   550'.222   81-21018

British Library Cataloguing in Publication Data
Roberts, John L.
Introduction to geological maps and structures.
—(Pergamon international library)
1. Geology—Maps 2. Geology, Structural
I. Title
551.8   QE601

ISBN 0-08-023982-X (Hardcover)
ISBN 0-08-020920-3 (Flexicover)

Printed in Great Britain by A. Wheaton & Co. Ltd., Exeter

Preface

THE elementary student of geology may well first encounter the basic elements of structural geology through the study and interpretation of geological maps. However, there are few if any textbooks on structural geology which concentrate on showing how the forms of geological structures can be determined from the evidence presented by a geological map, even although such an analysis is fundamental to their interpretation. Admittedly, there are several laboratory manuals on the British market which offer a very elementary introduction to the basic methods used in the interpretation of geological maps. These manuals are mostly based on so-called "problem maps", often of a rather artificial nature and usually based on a very simplistic view of the complexity shown by geological structures.

By way of contrast, the present book attempts a reasonably comprehensive account of geological structures and how they may be recognised through the study of geological maps, starting from first principles. It therefore provides an introduction to the basic methods used in the interpretation of geological maps, while it also attempts to cover the backround knowledge needed for a full understanding of geological maps. Any discussion of the mechanical principles involved in the formation of geological structures is kept to a minimum, since this aspect of the subject is covered adequately by the standard textbooks on structural geology.

The bulk of the text was written during the tenure of a Leverhulme Trust Fellowship which I acknowledge with much gratitude. I should also like to thank Christine Cochrane, who prepared all the illustrations from my very rough drafts, Elizabeth Walton and Sandra Elcock, who provided secretarial help at many stages in the preparation of the manuscript, and my colleagues in the Department of Geology at Newcastle who took an interest in the progress of the book from its inception, particularly Colin Scrutton who provided useful advice on stratigraphic principles and nomenclature.

Contents

CHAPTER 1

Introduction

Nature of Geological Maps

GEOLOGICAL maps are generally prepared in the field using a topographic map as a base. Such a base map, produced in Great Britain by the Ordance Survey, shows the form or topography of the land surface by means of topographic contours. These are drawn as lines of equal height above a reference plane, such as mean sea-level, which is known as the Ordnance Datum. Topographic contours are usually drawn at a particular interval, so that the contour height is a multiple of the contour interval. The topography of the sea floor can likewise be represented by submarine contours.

It is important to be able to visualise the form of the earth's surface by studying the contour pattern drawn on a topographic map (see Fig. 1.1).

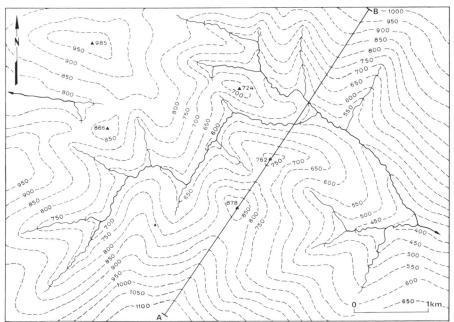

FIG. 1.1. Topographic map showing the form of the land surface by means of topographic contours drawn at an interval of 50 metres.

1

Particular features can be recognised from the contour pattern, thus allowing a mental picture of the topography to be constructed. Studying the natural features such as rivers, lakes, marshes, screes, cliffs and coasts, which are also shown in topographic maps, helps to identify the hills and valleys which together form the topography of an area.

The topographic map used as a base in geological mapping has a particular scale. This can be given in writing (1 inch equals 1 mile), as a fractional scale (1:63,360), or as a line printed on the map and divided into segments corresponding to certain distances on the ground.

Geological mapping is commonly undertaken at one scale, to be published at a different scale. For example, the Institute of Geological Sciences in Great Britain now undertakes mapping in the field at a scale of 10 centimetres to 1 kilometre (or 1:10,000), which is the metric equivalent of the old 6 inches to 1 mile scale, while it publishes maps on the following scales:

1:10,560	6 inches to 1 mile
1:25,000*	4 centimetres to 1 kilometre
1:50,000*	2 centimetres to 1 kilometre
or	
1:63,360	1 inch to 1 mile
1:100,000*	1 centimetre to 1 kilometre
1:250,000*	1 centimetre to 2.5 kilometres
or	
1:253,440	1 inch to 4 miles
1:625,000*	1 inch to about 10 miles
1:1,584,000	1 inch to 25 miles
1:2,500,000	1 centimetre to 25 kilometres

It is the maps on a metric scale, as marked by asterisks, which are gradually replacing the older maps on a similar scale, which were based on imperial measurements.

Although these maps become less detailed as the scale decreases, they are all based on the same information, as obtained from field mapping. It is the geological maps now being published on a scale of 1:50,000 (previously 1:63,360) which form a series of reasonably detailed maps covering the whole country. On a larger scale, the 1:25,000 maps depict areas of particular geological interest, while some 1:10,560 maps are published for certain areas, mainly coalfields. On a smaller scale, the 1:250,000 maps are intended to provide a complete coverage for the country without showing very much detail, while the geological maps published on the scale of 1:625,000 or 1:1,584,000 are simplified maps covering the whole country as two sheets (1:625,000) or a single sheet (1:1,584,000). The Geologic Quadrangle Maps produced by the United States Geological Survey on a scale of 1:24,000 or

1:62,500 are equivalent to the 1:50,000 maps for Great Britain, while geological maps are produced on a scale of 1:2,500,000 as four sheets covering the whole country. Geological maps on various other scales are also published for paticular purposes by the United States Geological Survey and by the corresponding organisations in the individual States.

Solid and Drift Maps.

Geological maps show the distribution of the superficial deposits and the underlying solid rocks at the earth's surface. The superficial deposits include glacial sands and gravels, boulder clay, alluvial sands and gravels, lacustrine silts and clays, screes and landslip deposits, beach sands and conglomerates, and peat. Such deposits form a thin and discontinuous mantle of material, covering the underlying rocks of the earth's crust. This material has been formed during the present cycle of erosion, and the deposits are mostly Pleistocene or Recent in age. This means that they have been formed during the last million years or so.

The geological maps published by the Institute of Geological Sciences for the various parts of Great Britain are produced in two editions. The Solid and Drift Edition shows the distribution of the superficial deposits by means of colouring. The solid rocks are only coloured on these maps where they are not overlain by superficial deposits. The older maps of this type were termed drift maps, simply because most of the superficial deposits in Great Britain are formed by glacial drift. The Solid Edition of the same maps does not show the superficial deposits by means of colouring. Instead, the solid rocks are coloured even where they are overlain by superficial deposits. Symbols printed on the map are used to show the nature of the superficial deposits. However, such maps often show the more widespread areas of river alluvium and wind-blown sand by means of colouring.

The Geologic Quadrangle Maps published by the United States Geological Survey mostly correspond to the Solid and Drift Edition of the geological maps produced by the Institute of Geological Sciences for Great Britain, since the superficial deposits are shown on these maps by means of colouring. However, some maps produced for areas covered by glacial deposits are published in a form similar to the Solid Edition.

Geological Boundaries.

The surface distribution of the solid rocks in a particular region is shown by means of geological boundaries, drawn as lines on the geological map (see Fig. 1.2). These boundaries define the outcrops of geological formations which have been mapped as separate entities in the field. A geological formation is simply defined as any body of rock which is sufficiently distinct that its boundaries can be mapped in this way.

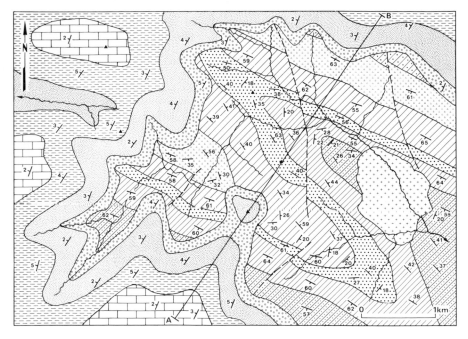

FIG. 1.2. Geological map depicting the outcrop of the geological formations recognised within the area of Figure 1.1. The ornament assigned to each geological formation is shown by the stratigraphic column of Figure 1.3.

The outcrop of a geological formation corresponds to the area which is directly underlain by rocks belonging to the formation. The outcrops of all the formations recognised in a particular region combine to form the outcrop pattern shown by a geological map. Exposures are only found where the underlying rocks are not hidden from view by soil or superficial deposits, so that they can be seen at the earth's surface.

Although the outcrop of a geological formation is mapped at the earth's surface, the formation is itself a three-dimensional body of rock. It has a particular shape, extending to some depth below the earth's surface, as defined by its contacts with the adjacent formations. It is very important to realise that such a formation, now outcropping at the earth's surface, originally extended to some height above the present level of this surface, before it was removed by erosion.

The accuracy of the boundaries shown on a geological map depends on two factors. Firstly, boundaries can only be drawn where there is a sharp contact between adjacent formations. Such boundaries become easier to distinguish as the rocks on either side of the contact become increasingly distinct. A boundary cannot be drawn with any degree of accuracy if it represents a transitional contact across which the rocks change only

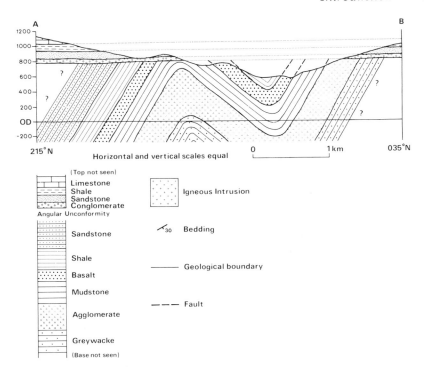

FIG. 1.3. *Above:* a vertical cross-section drawn along the line AB of Figure 1.2 to show the geological structure of the area. *Below:* stratigraphic column showing in chronological order the sequence of geological formations recognised within the area of Figure 1.2, with older rocks arranged below younger rocks.

gradually in character. Secondly, the ability to follow geological boundaries in the field depends on the degree of exposure, according to which the solid rocks tend to be hidden under a cover of soil and superficial deposits. It is common practice to show geological boundaries as solid or dashed lines, depending on whether or not these boundaries have been accurately located in the field.

The detail shown by a geological map depends primarily on the diversity of the rocks which have been mapped, since it is only variations in the nature of the rocks which allow different formations to be mapped. However, it also depends on the scale of the map and the degree of exposure. Geological maps showing areas of economic interest are often very detailed since the information obtained by mapping at the surface is supplemented by subsurface information which has been obtained from bore holes and mine workings.

Sections and Symbols.

Geological maps are generally accompanied by vertical cross-sections, a stratigraphic column, a legend or key showing the structural symbols, ornament and colours used on the map, and a brief description of the geology.

Vertical cross-sections are drawn across the map to illustrate the geological structure of the rocks, as shown in Figure 1.3. The lines of section are generally marked on the map. Each cross-section provides a vertical section through the upper levels of the earth's crust to show the geological structure extrapolated from the surface to a depth of a few kilometres. It should also be extended above the earth's surface to show the inferred nature of the geological structures above the present level of erosion. Such a cross-section is usually drawn at the same scale as the geological map. The vertical scale should correspond to the horizontal scale so that there is no vertical exaggeration to distort the form of the geological structures. The cross-section is then known as true-to-scale. It is constructed from the geological observations made at the surface, supplemented by subsurface data from mine workings and bore holes. It is mistakenly termed a horizontal cross-section on the maps published by the Institute of Geological Sciences in Great Britain because it is drawn along a horizontal line.

The stratigraphic column shows a generalised sequence of all the geological formations which have been mapped, arranged in chronological order from the oldest at the bottom to the youngest at the top, as shown in Figure 1.3. It is commonly drawn to scale, so allowing the thickness of each formation to be shown as appropriate. The symbol, ornament and colour assigned to each formation is shown by the stratigraphic column. It is termed a vertical section on the maps published by the Institute of Geological Science in Great Britain, since it corresponds to the sequence of formations which would generally be encountered with increasing depth. The stratigraphic column is often annotated with brief descriptions of the various formations. This annotation may be expanded to provide a short description of the geology, printed on the map. Alternatively, an explanatory memoir may be published seperately.

CHAPTER 2

Sedimentary Rocks and the Outcrop Pattern

Bedding of Sedimentary Rocks

SEDIMENTARY rocks are formed by material, mostly derived from the weathering and erosion of pre-existing rocks, which is deposited as sediment on the earth's surface. There are two main classes of sedimentary rocks, which differ in their mode of origin.

The clastic or detrital sediments are represented by rocks such as conglomerates, sandstones and shales. They are composed of rock fragments, mineral grains and clay minerals, produced by the physical and chemical weathering of older rocks. These particles are transported from their place of origin by the flow of air, water and ice across the earth's surface, accumulating as sedimentary deposits once the flow of the transporting medium so declines that the particles cannot be moved any farther. Deposition occurs progressively since the larger and heavier particles require more energy to be transposed in comparison with the smaller and lighter particles. This results in a sorting of the particles to form deposits of decreasing grain size away from their source.

The chemical and organic sediments comprise such rocks as limestones, dolomites, cherts, evaporites and coal. These rocks are mostly formed from material produced in solution by chemical weathering, which accumulates in sea water as the result of evaporation. This material in solution may be abstracted by direct precipitation to form chemical sediment, or it may be removed by organisms to form shells and skeletons, which then accumulate as organic sediment.

Beds and Bedding Planes.

Sedimentary rocks generally occur in the form of layers which were deposited on top of one another to form a stratified sequence as shown in Figure 2.1. Each layer is known as a bed or stratum (pl: strata), while the layered nature of sedimentary rocks which results from such a mode of formation is termed bedding or stratification. Sedimentary beds usually differ from one another in lithology, which is a general term referring to the overall character of a rock, as seen particularly in the field. The lithological

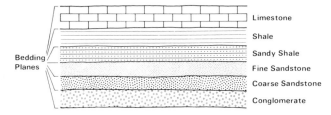

FIG. 2.1. Diagram showing the bedding or stratification of sedimentary rocks with beds of different lithology (conglomerate, sandstone, sandy shale and limestone) separated from one another by bedding planes. Each bed forms a parallel-sided or tabular layer. Such layers often divided internally by bedding planes to form a whole series of beds, differing slightly from one another in lithology.

differences reflected in the bedding of sedimentary rocks are usually defined by differences in mineral composition, grain size, texture, colour, hardness and so on, which allow the individual beds of sedimentary rock to be recognised.

The surfaces separating the individual beds of sedimentary rock from one another are known as bedding planes. They are commonly defined by a more-or-less abrupt change in lithological character which occurs in passing from one bed to the next. However, they may simply be defined by planes of physical discontinuity between beds of otherwise similar lithology, along which the rock will split.

The thickness of individual beds generally varies from several centimetres to a few metres, while their lateral extent may be measured in kilometres. Accordingly, beds of sedimentary rock typically form thin but very widespread layers, which eventually disappear as their thickness decreases to zero. Such beds may retain a lithological identity, or they may show a gradual change in lithological character, as they are traced throughout their outcrop. Some deposits of sedimentary rock show rapid changes in thickness so that they do not have such a parallel-sided form. For example, elongate bodies of sandstone and conglomerate are laid down in river channels and along shore lines, wedge-shaped bodies of breccia and conglomerate are banked against buried hills, volcanoes and coral reefs, while the dome-shaped form of reef knolls and the elongate form of barrier or fringing reefs are the result of organic activity. However, such examples are exceptions to the general rule that most beds of sedimentary rocks occur effectively in a tabular form as parallel-sided layers which extend laterally for a considerable distance before they eventually disappear as their thickness decreases to zero.

A bedding plane generally represents the original surface on which the overlying bed of sedimentary rock was deposited. Such a surface is simply known as a surface of deposition. Since sedimentary rocks are deposited on the earth's surface, such a bedding plane corresponds to this surface just before the overlying rocks were laid down. Commonly, the earth's surface is very close to the horizontal wherever sedimentary rocks are accummulating as the result of deposition at the present day. For example, the alluvial plains and deltas formed by the lower reaches of large rivers, and the shallow seas into which they flow beyond the land, are typically flat-lying areas where considerable thicknesses of sedimentary rocks have just been deposited. However, since a bedding plane is considered to represent such a surface of sedimentary deposition, it can be argued by analogy with the present day that this surface was virtually horizontal when the overlying bed of sedimentary rock was deposited.

This implies that sedimentary rocks were originally deposited as nearly if not quite horizontal beds. Accordingly, sedimentary rocks are generally deposited with what is known as a low initial dip, unless they were deposited on a sloping surface. Such exceptions are only likely where sedimentary rocks are deposited on the flanks of deltas, coral reefs, volcanoes and hill slopes. The beds rarely have an initial dip greater than 30° from the horizontal, even under these circumstances. They are usually found to flatten out as they are traced away from such prominences, until they eventually become horizontal or nearly so.

Stratigraphic Order and Superposition.

Stratigraphy can be defined as that branch of geology which studies the stratified nature of sedimentary rocks as a basis for interpreting the geological history of the earth's crust. The study of stratigraphy starts with establishing the chronological order in which a sequence of sedimentary rocks was originally deposited at the earth's surface. This is known as the stratigraphic order of the sedimentary rocks. What is known as a stratigraphic sequence can be established by placing the sedimentary rocks in their chronological order.

The stratigraphic order of sedimentary rocks can simply be established according to the Principle of Superposition which states that, while each bed in a stratified sequence is younger than the underlying bed, it is older than the overlying bed. The principle arises simply because sedimentary rocks are laid down as flat-lying beds on top of one another, so that they become progressively younger in an upward direction. In other words, the oldest rocks occur at the base of a stratigraphic sequence, while the youngest beds are found at the top. The stratigraphic order of sedimentary rocks can be shown in this way by means of a stratigraphic column, such as commonly accompanies a geological map (See Fig. 1.3).

Sedimentary rocks are commonly if not invariably found to be affected by earth movements in such a way that the bedding is no longer horizontal. For example, they may become tilted so that the bedding is inclined in a particular direction. This is known as the direction of dip, and the bedding is said to dip at a particular angle from the horizontal in this direction, which corresponds to the dip of the sedimentary rocks. The horizontal direction of right angles to the dip of the bedding is known as the strike. How the dip and strike of a bedding plane can be measured in the field will be described later in this chapter (See Fig. 2.4). The vertical cross-section of Figure 1.3 shows the upper sequence of sedimentary rocks to be very slightly tilted, since the bedding dips at a low angle to the west-north-west. Alternatively, sedimentary rocks may become folded so that the bedding is thrown into a wave-like form, as shown by the lower sequence of sedimentary rocks in the vertical cross-section of Figure 1.3. The bedding shows regular changes in dip as it is traced across these folds in the sedimentary rocks.

Accordingly, earth movements can affect sedimentary rocks in such a way that the bedding becomes increasingly inclined away from the horizontal until it reaches a vertical position, beyond which it may become overturned. The beds are then said to be inverted, even although the bedding may still be inclined away from the horizontal. The Principle of Superposition does not apply to sedimentary rocks which have been so affected by subsequent earth movements that they are now inverted. It is often difficult to determine the stratigraphic order of sedimentary rocks in such regions. However, it may be possible to trace these rocks into areas of less complexity, where the stratigraphic order can be determined according to the Principle of Superposition.

Stratigraphy of Sedimentary Rocks

Sedimentary rocks often contain fossils as the organic remains of plants and animals, which were incorporated into the sediment at the time of its deposition. It was early recognised that the fossils preserved in a sedimentary rock are diagnostic of its geological age, unless they have been derived from the erosion of older rocks. This is simply a consequence of organic evolution whereby certain species change into new and different forms, combined with organic extinction whereby other species die out completely, with the passing of geological time. Thus, each species is only present as a living organism during a particular time span, before which it had not evolved from a pre-existing form and after which it had either become extinct or evolved into a new form. This means that sedimentary rocks can be dated as belonging to a particular interval of geological time, according to the assemblage of fossil species which they contain. This principle allows stratigraphic correlation since it implies that sedimentary rocks containing similar assemblages of fossil species are the same age.

The Stratigraphic Time-scale.

The evolution and extinction of organisms gives a definite and recognisable order to the succession of fossil assemblages which are preserved in sedimentary rocks. This means that the fossil record can be used to erect a stratigraphic time-scale which allows geological time to be divided into distinct units. Arranged in order of decreasing length, these units are known as Eras, Periods, Epochs and Ages. These terms refer to particular intervals of geological time as defined by the fossil record. Thus, geological history over approximately the last 560 million years is divided into the Cenozoic, Mesozoic and Palaeozoic Eras. Each of these eras is divided into a number of periods, as shown in Table 2.1. Each period is formed by a number of epochs, and this division of geological time can be continued by dividing each epoch into a number of different ages. However, it is rarely possible to recognise such a short interval of geological time on a world-wide basis. The table shows that the major divisions of the stratigraphic time-scale can be dated in millions of years using radiometric methods. The dates given are those shown in the 1976 edition of the International Stratigraphic Guide.

Chronostratigraphic Divisions

The sedimentary rocks deposited during a particular interval of geological time form a series of stratigraphic units which correspond to the eras, periods, epochs and ages of the stratigraphic time-scale as follows:

Era	----
Period	System
Epoch	Series
Age	Stage

The terms on the left constitute a hierarchy of geological time units (or geochronological units) which is used to divide up geological time, while the terms on the right represent the equivalent hierarchy of time-stratigraphic units (or chronostratigraphic units), which is used to divide up the stratigraphic record according to geological age. These time-stratigraphic units are named after the particular intervals of geological time when they were deposited. Thus, the rocks of the Tertiary System were deposited during the Tertiary Period. They are formed by the Pliocene, Miocene, Oligocene, Eocene and Palaeocene Series, which were deposited during the corresponding epochs of the Tertiary Period (see Table 2.1). Since epochs and series often have no specific names, individual periods are often divided into early, middle and late epochs, while the corresponding parts of the equivalent system are termed the lower, middle and upper series of that system. For example, the rocks of the Lower Carboniferous Series were deposited during the Early Carboniferous Epoch.

The rocks of a stratigraphic series can be further divided into stages, characterised by fossil assemblages which are considered to be diagnostic of

TABLE 2.1

The Stratigraphic Time-scale and Corresponding Chronostratigraphic Units

Era	Period or system		Epoch or series		Age (my)
CENOZOIC	Q QUATERNARY	* *	Recent Pleistocene		
					2 —
	T TERTIARY	k^1 **	Pliocene Miocene	NEOGENE	
		i^{8-12} i^{3-7} i^{1-2}	Oligocene Eocene Palaeocene	PALAEOGENE	
					64 —
MESOZOIC	K CRETACEOUS	h^5 h^{1-4}	Upper Lower		
					140 —
	J JURASSIC	g^{10-14} g^{5-9} g^{1-4}	Upper (Malm) Middle (Dogger) Lower (Lias)		
					208 —
	℞ TRIASSIC	f^{4-6} ** f^{1-3}	Upper (Keuper and Rhaetic) Middle (Muschelkalk) Lower (Bunter)		
					242 —
PALAEOZOIC	P PERMIAN	e^{1-5} ** **	Upper Middle Lower		
					284 —
LATE	CARBONIFEROUS	d^{4-6}	Upper	∼ P PENNSYLVANIAN	
		d^{1-2}	Lower	∼ M MISSISSIPPIAN	
					360 —
	D DEVONIAN	c^3 c^2 c^1	Upper Middle Lower		
					409 —
PALAEOZOIC	S SILURIAN	b^7 b^6 b^5	Upper (Ludlow) Middle (Wenlock) Lower (Llandovery)		
					436 —
EARLY	O ORDOVICIAN	b^{3-4} b^{1-2}	Upper (Caradoc and Ashgill) Lower (Arenig, Llanvirn and Llandeilo)		
					500 —
	Є CAMBRIAN	a^3 a^2 a^1	Upper (including Tremadoc) Middle Lower		
					564 —
PRECAMBRIAN	pЄ				

* Quaternary deposits are represented by symbols which indicate their lithological nature.

**These chronostratigraphic units are not represented in the geological record of the British Isles.

Note: There is generally no d^3 horizon shown on the geological maps of the British Isles, as the result of an historical accident.

their geological age. However, it is the longer intervals of geological time, and the corresponding parts of the stratigraphic record, which mostly concern us in the analysis of geological maps in terms of geological history.

Mapping of Sedimentary Rocks.

A stratigraphic formation is simply defined as any body of sedimentary rock which is sufficiently distinct that its boundaries can be mapped in the field. Most formations occur as sheet-like bodies with a horizontal extent which is very much greater than their vertical thickness. It is commonly found that such formations vary in thickness from a few metres to several thousand metres, while it may be possible to trace individual formations for several hundred kilometres. The boundaries between stratigraphic formations usually appear to be parallel to the bedding of the underlying rocks. Such a boundary is said to be conformable wherever this is the case. This means that stratigraphic sequences tend to accumulate as vertical successions of widespread but flat-lying formations, which are conformable with one another throughout their outcrop.

Since the formation is the basic unit in the mapping of sedimentary rocks, it must have a lithological character which allows it to be distinguished from adjacent formations. This character can be defined in various ways. Some formations may contain beds of only a single lithology, while others may be formed by alternating beds of more than one lithology. A formation may be characterised by a particular lithology, even if more than one lithology is present, or by a particular association of different lithologies, which are often arranged in the form of rhythmic or cyclic sequences. Occasionally, formations have a lithological diversity which is itself a distinctive characteristic. The stratigraphic columns accompanying the geological maps of the United States Geological Survey show many of the features which serve to define particular formations as mappable units in the field.

Formations are generally give a geographical name, followed by a term descriptive of the dominant lithology (Loch Tay Limestone Formation). However, if a formation is composed of more than one lithology, it is simply called a formation without any term descriptive of the lithology (Caseyville Formation).

Lithostratigraphic Divisions

Individual formations may be divided into members if there is any part of the formation which is sufficiently distinct to be mapped as a local unit. Several members may be recognised in a formation, together constituting the formation, or a single member may be identified as an isolated unit within a formation, which is otherwise not divided into parts. Finally, distinctive parts of a formation which are laterally equivalent to one another may also be considered as members of the formation. Members are named in a similar way to formations, except that a term descriptive of the lithological character is nearly always used (Big Clifty Sandstone Member of the Golcanda Formation).

It is occasionally possible to trace individual beds of sedimentary rock throughout their outcrop if they have some uniquely distinguishing feature. Such key beds may be identified by their lithological character or fossil content. For example, bentonite beds are formed by a particular type of clay which is produced by the alteration of volcanic dust under marine conditions. Such beds can be traced for considerable distances if they occur so rarely that individual beds can be recognised within the sequence. Likewise, a fossil may serve to identify a particular bed if it does not occur elsewhere in the succession. However, the absence of such distinguishing features usually means that individual beds cannot be traced from exposure to exposure with any degree of certainty.

While formations may be divided into members and beds, they may also be combined to form groups and supergroups, defined on the basis of the lithological characteristics which they share in common with one another. There is, therefore, a hierarchy of stratigraphic rock units arranged in order of decreasing size:

Supergroup
Group
Formation
Member
Bed

which is recognised in the mapping of sedimentary rocks. The rock units forming this hierarchy are known as rock-stratigraphic units (or lithostratigraphic units) in order to distinguish them from time-stratigraphic units. It is the boundaries between these rock-stratigraphic units, forming the rock bodies that can be mapped in the field, which are shown on a geological map.

Lateral Variations in Sedimentary Rocks.

Although rocks of more than one lithology are commonly found within a stratigraphic formation, sedimentological and fossil evidence usually suggests that they were laid down in the same depositional environment. It is a matter of common observation that depositional environments vary from point to point on the earth's surface, and that the distribution of these environments is reflected by changes in the lithology of the sediments being deposited, at the present day. However, the evidence from the geological record indicates that the deposition of sedimentary rocks also varies with the passing of geological time. This is simply shown by the vertical succession of stratigraphic formations, each formed by sediments differing in lithology, which were deposited during successive intervals of geological time at any one locality.

Now consider a formation boundary which is marked by a lithological change from sandstone to shale. This obviously reflects a change in the lithology of the sediment being deposited. This change may take place at practically the same time throughout the region where the formations have been mapped. This would mean in effect that the formation boundary is a bedding plane representing the depositional surface on which the lowermost bed of shale was laid down. However, it is equally possible that the change from sandstone to shale did not occur at the same time throughout the whole area where the two formations have been mapped. This would mean that sandstone was being deposited in one area, while shale was being deposited in another area, at any particular time. Individual beds of sandstone might disappear as their thickness gradually decreased to zero, or they might show a lateral change in lithology from sandstone to shale. In either case, the formation boundary would be slightly oblique to the bedding of the rocks, assuming that the bedding planes represent a vertical succession of depositional surfaces. Such a formation boundary is said to be diachronous. Such boundaries are developed wherever sedimentary rocks show lateral changes in lithology between two formations.

Distinction between Chronostratigraphy and Lithostratigraphy.

It is likely that most boundaries between stratigraphic formations are diachronous to some extent. This means that a clear distinction must be made between chronostratigraphic units defined in terms of geological time, such as systems, series and stages, and lithostratigraphic units defined in terms of lithology, such as groups, formations and members. Accordingly, sedimentary rocks may be divided on fossil evidence into systems, series and stages to form a hierarchy of chronostratigraphic units corresponding to the various intervals of the stratigraphic time-scale. The basis for this division is the recognition of biostratigraphic zones characterised by particular assemblages of fossil species. Likewise, geological mapping allows the sedimentary rocks of a particular region to be divided into groups, formations and members. The vertical sequence shown by these divisions of the stratigraphic column, and any lateral changes recognised in these divisions as they are traced throughout their outcrop, defines the lithostratigraphy of the particular region under consideration.

This distinction between chronostratigraphic and lithostratigraphic units is clear wherever formation boundaries can be shown to transgress biostratigraphic zones along their outcrop (See Fig. 2.2). However, it is rarely possible to show the boundaries of biostratigraphic zones on a

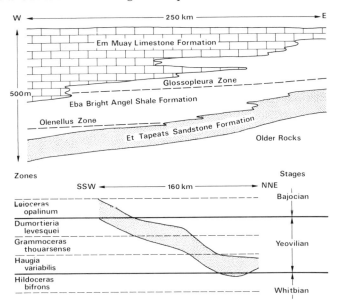

FIG. 2.2. *Above:* diagram showing how the boundaries of different lithostratigraphic formations in the Cambrian rocks of the Grand Canyon are cut across by biostratigraphic zones (after McKee). *Below:* diagram showing the diachronous nature of the Cotsworld-Bridport Sandstone Formation in Southern England (after Rayner). The fossil names on the left refer to the various zone fossils which are used to define particular stages of the Jurassic system, shown on the right. The vertical scale represents neither the thickness of the sandstone formation, nor the length of geological time corresponding to the various chronostratigraphic stages.

geological map for two reasons. Firstly, it must be recognised that the boundaries between biostratigraphic zones and stratigraphic formations are almost parallel to one aother. This means that diagrams like Figure 2.2 always have a vertical scale which is very much greater than the horizontal scale. This vertical exaggeration introduces the dimension of geological time into the diagram, thereby allowing the diachronous nature of the formation boundaries to be distinguished. Such a cross-section can best be termed a stratigraphic cross-section to distinguish it from a structural cross-section drawn true to scale. Secondly, while geological boundaries can easily be traced between different formations in the field, it would be very difficult to trace the boundaries of biostratigraphic zones in the same manner. Such boundaries are usually based on the detailed study of the fossil record at a series of separate localities, between which they are drawn by interpolation.

This means that geological maps rarely show the fine detail of stratigraphic relationships, so placing a strict limit on what can be deduced from their study in terms of geological history. It is often very difficult if not impossible to determine any lateral changes in the nature of the sedimentary rocks being deposited at a particular time from the evidence shown by a geological map. Such changes can usually only be recognised if individual lithostratigraphic units disappear as they are traced along their outcrop. It should be clearly recognised that the distinction made between lithostratigraphic and chronostratigraphic units does imply that sedimentary formations are likely to have diachronous boundaries. However, if this is the case, it is still possible to determine the major changes which occurred during the deposition of a sedimentary sequence from the superposition of the stratigraphic formations in a particular order. It is this evidence which is most clearly shown by a geological map.

Nature of the Stratigraphic Record.

We can now consider the methods used in the interpretation of geological maps. These methods are concerned with determining the geological structure in three dimensions from a study of the outcrop pattern and the structural observations plotted on a geological map. This is known as reading a geological map. The geological structure is defined by the form and mutual arrangement of the geological formations below the earth's surface, while the outcrop pattern is defined by the boundaries which are shown on a geological map between these formations. It will first be assumed that these boundaries represent the contacts between stratigraphic formations, which generally occur in the form of parallel-sided layers of a constant thickness, arranged as a conformable sequence of sedimentary rocks.

Establishing the Stratigraphic Sequence.

Although the outcrop patterns developed by sedimentary rocks are very diverse, they all show one feature in common. Thus, each formation in a conformable sequence of sedimentary rocks is always flanked by two other formations. One of these formations occurs stratigraphically below the particular formation in question, so that it is older than this formation, while the other formation occurs stratigraphically above this formation, so that it is younger. The two-sided nature of stratigraphic formations is clearly shown by the outcrop pattern of sedimentary rocks which have been tilted or folded. Each formation can then be traced as a relatively narrow outcrop across the map, sandwiched between the two other formations, as shown for example, in Figure 1.2. However, the outcrop pattern of flat-lying formations also shows that each formation only occurs in contact with two other formations even although the outcrop pattern, at least in areas of high topographic relief, is often rather complex. Such an arrangement is only modified if a particular formation wedges out along its outcrop, so that the two other formations on either side come into contact with one another, while the missing formation forms what is known as a feather edge at its point of disappearance.

This means that it is usually possible to arrange a conformable series of sedimentary formations into a sequence which either corresponds to the stratigraphic order, or its reverse. This is done simply by examining the outcrop pattern to determine which formations are seen in contact with one another. For example, if formation A is seen in contact with formation B, while formation B is also seen in contact with formation C, the stratigraphic sequence must be ABC or its reverse, depending on whether formations A and B are older or younger than formations B and C respectively. Once the contact relationships of all the sedimentary formations shown by the geological map have been examined, it should be possible to draw up a complete stratigraphic sequence in this way.

Stratigraphic Order of Sedimentary Formations.

Once a stratigraphic sequence has been established for the sedimentary formations shown by a geological map, it is necessary to determine its stratigraphic order. This can be done directly if there is palaeontological evidence to date the sedimentary formations according to the stratigraphic time-scale. Such evidence is usually shown by the stratigraphic column which normally accompanies a geological map. This will show the sedimentary formations arranged in their correct order from the oldest formation at the base to the youngest formation at the top. This is how any stratigraphic sequence should be given. If such a stratigraphic column is not available, the stratigraphic order can be determined according to the Principle of Superposition from the structural observations plotted on the geological map to show the dip of the bedding or from the effect of the topography on the outcrop of the geological boundaries, as described later in this chapter.

The geological maps published by the United States Geological Survey show the stratigraphic age of each formation by means of a capital letter, which refers the formation to a particular system. These capital letters are listed in front of the system names given in the second column of Table 2.1, except that the Carboniferous System is divided into Mississippian (M) and Pennsylvanian (P1). The letters in the lower casement following the capital letter simply give the initials of the formation name (or its lithological character in the case of Quaternary deposits and igneous rocks).

A similar method is used by the Institute of Geological Sciences in Great Britain, as shown by the symbols given in front of the epoch names listed in the third column of Table 2.1. The individual symbols (e.g. g^7) refer to particular formations or stratigraphic groups. However, the initial letter gives the system to which these rock-stratigraphic units belong, while the numerical superscript gives the approximate age of these lithostratigraphic units in terms of the chronostratigraphy. The single asterisk in the third column of Table 2.1 means that no letter is allocated to Pleistocene and Recent deposits. Instead, abstract symbols are used to show the lithological nature of these superficial deposits. The double asterisks shown in the third column of Table 2.1 indicate the time-stratigraphic units which are missing from the stratigraphic record of Great Britain. Finally, it should be noted that there is usually no d^3 horizon, as the result of an historical accident. This system obviously does not allow a very clear distinction to be drawn between lithostratigraphic and chronostratigraphic units.

Lateral Changes in Stratigraphic Sequence.

It is commonly found that the same sequence of sedimentary formations is present throughout the whole area of a geological map. However, this is not always the case, since individual formations may disappear as they are traced in a particular direction. This may occur at the base of a stratigraphic sequence, wherever an individual formation disappears by wedging-out against an area of older rocks. Alternatively, individual formations within a stratigraphic sequence may thin to zero thickness along their outcrop, so allowing the formations on either side to come into contact with one another. The disappearance of stratigraphic formations in this way is simply the result of lateral changes in the stratigraphic sequence. Such formations are usually shown as wedge-shaped bodies in the stratigraphic column accompanying the geological map.

There is a particularly useful method which allows such lateral variations in the stratigraphic sequence to be visualised. The geological map should be viewed in an oblique direction which corresponds to the dip of the sedimentary formations in a particular area. Although it is not strictly necessary, the geological map can be held so that the line of sight is horizontal. This provides a foreshortened view of the outcrop pattern. It is equivalent to drawing a cross-section parallel to the strike in such a way that it is perpendicular to the dip of the bedding. Such a down-dip view of the outcrop pattern shows the true thickness of the sedimentary formations, which appear to be horizontal. It then corresponds to a stratigraphic cross-section showing how the various formations disappear along their outcrop. If these formations are folded, it would be possible to visualise a composite cross-section formed by changing the down-dip view of the outcrop pattern as the bedding varies in attitude across the area. This down-dip method of viewing the outcrop pattern has several other applications in the study of geological maps. It can only be used where the topography has little or no effect on the outcrop of the geological boundaries.

Gaps in Stratigraphic Sequence.

Faults are mapped as a distinct type of geological boundary, representing a fracture across which the rocks on either side have been displaced relative to one another. This introduces a structural discontinuity into the outcrop pattern, across which the geological boundaries between the stratigraphic formations are displaced, as shown by the geological map of Figure 1.2. If this displacement is sufficiently large, it may not be possible to match the individual formations across the fault, simply because they are not exposed on both sides. For example, formation A and B might be exposed on one side while formations Y and Z outcrop on the other side.

It would then be imposible to determine the stratigraphic relationships between the two sequences, separated from one another by the fault, since formations A and B might be older or younger than formations Y and Z. However, these formations may be dated stratigraphically on palaeontological evidence, or there might be structural evidence to suggest how they should be placed in stratigraphic order. If this were the case, a stratigraphic column of sedimentary formations could be erected for the area under consideration. This should clearly show the position of such gaps in the stratigraphic record. The magnitude of such a gap can only be determined if it is possible to date stratigraphically the two formations which occur in faulted contact with one another. Otherwise, it would not be possible to determine how many formations are missing from the stratigraphic record.

Inliers and Outliers.

Once the stratigraphic order of the sedimentary formations has been established, inliers and outliers can be recognised as part of the outcrop pattern. An inlier is the outcrop of an older formation surrounded entirely by younger rocks, while an outlier is the outcrop of a younger formation surrounded entirely by older rocks, as illustrated by Figure 2.3. Inliers and outliers may simply result from the nature of the topography, or they may be developed as the result of folding and faulting.

Use of Structural Observations

Although the relative order of the sedimentary formations shown by a geological map can be determined simply by examining the stratigraphic column which usually accompanies the map, it is pedagogically more effective to disregard this source of information in learning to read a geological map. This can simply be done by cutting off the stratigraphic column so that only the evidence presented by the geological map itself can be used. The Geological Quadrangle Maps published by the United States Geological Survey are particularly useful for this purpose, since the sedimentary formations are only assigned the stratigraphic age of a System. The structural observations plotted on the geological map can then be used to determine the dip of the sedimentary formations, so allowing their stratigraphic order to be established according to the Principle of Superposition.

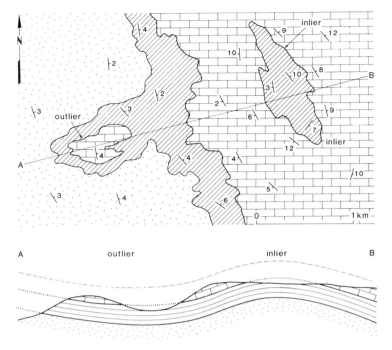

Fig. 2.3. Geological map and vertical cross-section showing the presence of an outlier of younger rocks and an inlier of older rocks.

Dip and Strike.

The inclination of the bedding can be defined by the dip and strike of a bedding plane. The dip measures the inclination of the plane away from the horizontal, while the strike refers to the trend of a horizontal line lying within the plane, relative to true north. Once measured in the field, these observations can be plotted on a geological map to give the attitude of the bedding at a particular point.

The strike is first determined by drawing a horizontal line on the exposed surface of a bedding plane in the field, using a clinometer as shown in Figure 2.4. The bearing of this line is then measured relative to magnetic north using a compass. This reading is corrected by adding or subtracting the magnetic declination to give the bearing from true north. This is usually measured clockwise within a 360° circle from true north: a south-westerly bearing would be given as 225°N. It should be noted that any surface has two direction of strike at 180° to one another. One convention gives that direction of strike from which the surface dips to the right. The direction of dip would then be clockwise from the direction of strike.

Although the dip can be given as a gradient, it is usually measured as an angle from the horizontal, as shown in Figure 2.4. Once the strike has been

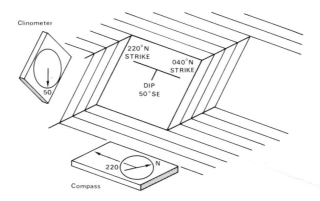

FIG. 2.4. Diagram showing how the attitude of a bedding plane is measured in the field to find its dip and strike.

determined, a line is drawn down the dip of the bedding plane, at right angles to the strike. The angle of this line from the horizontal is measured using a clinometer by placing it vertically along this line. This gives the angle of dip shown by the bedding plane. The horizontal trend of the line drawn down the dip of the bedding plane can then be measured with a compass, placed alongside the clinometer. Once corrected for the magnetic declination, this reading gives the direction of dip shown by the bedding plane, relative to true north. It should differ by 90° from the direction of strike measured for the same plane.

Once these measurements have been made in the field, there are two ways to describe the attitude of the plane. Firstly, the plane can be said to strike at 040°N, while it dips at 50° towards the south-east. The approximate direction of dip must be given in this case, so distinguishing planes with the same direction of strike but dipping in opposite directions from one another. For example, there is also a plane striking at 040°N which dips at 50° towards the north-west. This assumes that the strike has not been given strictly as that direction from which the plane dips towards the right. Secondly, the plane can be said to dip at 50° towards 130°N. The direction of dip must be given as accurate bearing from true north in this case. This method has the advantage that there is no uncertainty about the direction of dip, which is clearly specified as 130°N.

Such observations are commonly plotted on a geological map. This can be done using strike symbols or dip arrows. Strike symbols are drawn as a long bar parallel to the direction of strike, with a short tick on one side to indicated the direction of dip, as shown in Figure 1.2. Dip arrows are drawn at right angles to the direction of strike, pointing in the direction of dip. The angle of dip should be written alongside the symbol in both cases. The use of dip arrows is generally being superseded by strike symbols on geological maps which are now being published.

Since the bedding of sedimentary rocks is virtually parallel to the contacts of most stratigraphic formations, the observations of bedding plotted on a geological map can then be used to determine the attitude of these contacts. It may then be inferred from the Principle of Superposition that the sedimentary formations become younger in the direction of their dip, as shown by these observations, unless there is any evidence that the bedding is inverted. However, care is required because these observations come from individual exposures which may not be representative of the whole. For example, the bedding of a sedimentary formation may be disturbed locally by folding or faulting. This can result in anomalous dips, which are not the same as the overall dip of the formation, particularly in the vicinity of faults. There is also the possibility of local overturning. The stratigraphic order should therefore be determined wherever possible according to the Principle of Superposition so that local inversions can be recognised wherever the sedimentary formations appear to dip in the wrong direction. This method should obviously not be used if there is any possibility of widespread inversion, since it is then impossible to determine which rocks are inverted unless their stratigraphic order is already known. Overturned and inverted rocks are usually found in areas where there is clear evidence of intense structural disturbance.

Stratigraphic Use of Sedimentary Structures.

It has just been mentioned that the stratigraphic order of sedimentary formations cannot be determined according to the Principle of Superposition if there is any likelihood that the bedding is now inverted. It can be shown that sedimentary rocks are inverted if the stratigraphic column accompanying the geological map, becomes older rather than younger in the direction of their dip. However, if such evidence is lacking, there is a wide variety of sedimentary structures which can be used to distinguish the depositional base of a sedimentary bed from its top. This then allows the stratigraphic order of the beds to be determined in the field. The main types of sedimentary structures used in this way to determine whether the bedding is right-way-up or inverted are shown in Figure 2.5.

(a) Cross-bedding in which the foreset surface are concave upwards so that they are asymptotic to the base of the bed, while they are truncated by a surface of erosion at its top.

(b) Graded bedding in which there is a gradual decrease in the grain-size of the detrital particles from the base of the bed to its top. It should be noted, however, that reversed grading has been recorded, even although it appears to be very rare.

(c) Slump structures in which the folds and other structures affecting the slumped bed have been truncated by a surface of erosion at its top.

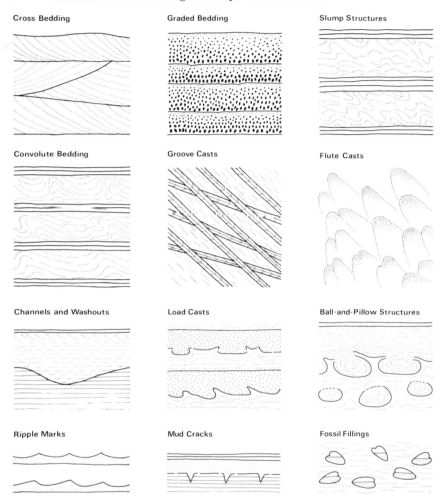

Cross Bedding

Graded Bedding

Slump Structures

Convolute Bedding

Groove Casts

Flute Casts

Channels and Washouts

Load Casts

Ball-and-Pillow Structures

Ripple Marks

Mud Cracks

Fossil Fillings

FIG. 2.5. Sedimentary structures used to determine the way-up of individual beds of sedimentary rock. See text for details.

(d) Convolute bedding in which the internal bedding planes are folded to form broad, rounded troughs separated from one another by narrow, cuspate crests.

(e) Sole structures formed by the differential erosion of the underlying bed, usually by currents responsible for the deposition of the overlying bed, or by the differential loading of the underlying bed by the weight of the overlying bed after deposition. The former category is represented by structures such as groove casts, flute casts, rill marks, sedimentary channels, wash-outs and bounce marks, while the latter category is represented by structures such as load casts and ball-and-pillow structures. All these structures are typically developed at the base of sandstone beds laid down in water on top of fine-grained muddy sediments. Even if such structures are not well developed, it is commonly found that beds of sandstone have upper surfaces which are much smoother to the touch than the lower surfaces.

(f) Ripple marks in which rounded troughs contrast with sharp crests. This form is typical of symmetrical or oscillatory ripples, but it is occasionally seen in asymmetrical or current ripples. It is often associated with a form of internal cross-bedding known as rippledrift bedding.

(g) Mud-cracks and sedimentary dykes which are infilled by material from the overlying bed.

Organic remains may also provide evidence concerning the stratigraphic order of the beds. Fossils may be preserved in their original positions of growth, or they may be reorientated after death into equilibrium positions by current action. Cavities within fossil remains may be partially filled with sediment from the base upwards. Trace fossils such as tracks, trails, castings, burrows and rootlets often have a characteristic form which can be recognised in the field.

The way-up of the individual beds forming the stratigraphic sequence can be determined as normal (right-way-up) or inverted wherever the depositional bases of the beds can be recognised using the structures just described. The beds are then said to young in that direction (normal to the strike) in which the beds become younger. This direction of younging corresponds to the direction of dip if the beds are right-way-up, whereas it is given by the opposite direction if the beds are inverted. The stratigraphic formations become younger in the opposite direction to the dip in the latter case, so allowing their stratigraphic order to be determined. Since the bedding within a sedimentary formation may be folded on a small scale, most reliance should be placed on observations which are close to the formation boundaries.

The structural observations plotted on a geological map commonly show by the use of special symbol if the beds are known to be inverted. This symbol would obviously not be used if there was any evidence that the beds were right-way-up. However, the normal dip-and-strike symbol is also used if there is no evidence of stratigraphic order. Accordingly, the use of a special symbol to show inverted bedding leads to some ambiguity about the meaning attached to the normal dip-and-strike symbol. It should be realised that this symbol would still be used for inverted bedding if there was no evidence from sedimentary structures to show that the bedding is inverted.

Width of Outcrop

Many formations are bounded by contacts which are parallel to one another, so forming a tabular or sheet-like body which has a constant thickness normal to its boundary surfaces. This thickness is termed the true or normal thickness T_n, in contrast to the apparent or vertical thickness T_Z which is measured in a vertical direction (see Fig. 2.6). The apparent

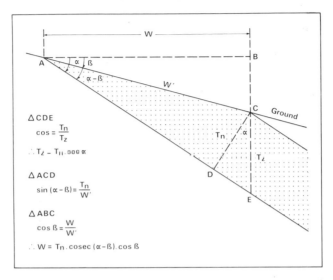

FIG. 2.6. Vertical cross-section illustrating how the width W of outcrop shown in a horizontal plane on a geological map varies with the dip α of the formation and the slope ß of the ground.

thickness T_Z increases in relation to the true thickness T_n as the dip of the formation increases from zero, according to the relationship:

$$T_Z = \sec \alpha \, T_n$$

where α is the angle of dip.

The dip of such a parallel-sided formation can be determined approximately by considering the width of its outcrop in relation to the slope of the ground. This is generally measured in the direction of its dip. The distance $W' = AC$ between the contacts of such a body gives the width of its outcrop, as measured parallel to the surface of the ground (see Fig. 2.6). However, this is shown on a geological map by the horizontal distance $W = AB$ between the geological boundaries defining the outcrop of such a formation. Simple trigonometry then indicates that the width of outcrop shown on a geological map is given by:

$$W = T_n \cosec (\alpha - \beta) \cos \beta$$

where the angle β gives the slope of the ground.

This means that the dip of a formation can be estimated from variations in the width of its outcrop, assuming that the effect of topography is taken into account. The relationship between width of outcrop and the dip of the formation is complex in areas of high relief. However, it is much simpler in areas of low relief, where the ground approaches the horizontal. The width of outcrop then decreases to a minimum value, corresponding to the true thickness T_n of the formation, as the dip of the formation increases to the vertical. This means that the outcrop of a formation becomes wider as its angle of dip approaches the horizontal. It should be noted that only the angle of dip can be determined in this way. The direction of dip must be found from other considerations, such as the effect of topography on the outcrop of the formation boundaries.

Topographic Expression of Geological Structure

Only the effect of topography on the nature of the outcrop pattern has been considered so far. However, the topography is often closely related to the outcrop pattern as a direct reflection of the geological structure of the underlying rocks. This is particularly the case if these rocks form a series of layers which differ in their resistance to weathering and erosion. The topographic expression of these differences varies according to the attitude of the layers, as shown in Figure 2.7.

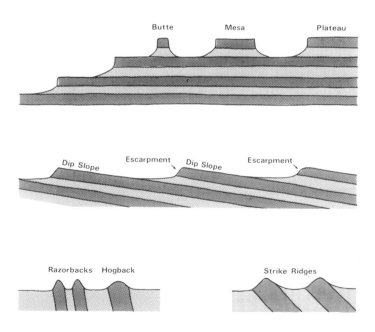

Fig. 2.7. Vertical cross-sections showing the typical landforms developed in response to the dip of the underlying rocks. The stippled layers are more resistant to weathering and erosion.

Flat-topped mesas and isolated buttes are formed wherever the layers are horizontal. The tops of these landforms are capped by a resistant layer, while their sides are formed by the underlying rocks which are less resistant to erosion. A cliff encircling the mesa or butte is often formed by the resistant layer. The sides of the landform have step-like breaks in slope wherever layers differing in their resistance to erosion are found in the underlying rocks.

Steep escarpments backed by gentle dip slopes are formed wherever the layers are inclined at a low angle. This topographic forms is known as a cuesta. The upper part of each escarpment, and the dip slope behind each escarpment, are formed by a layer resistant to erosion. The break of slope below the top of the escarpment marks the base of this layer. The lower part of the scarp slope, and the intervening valleys, are underlain by rocks which are less resistant to erosion. The escarpments are roughly parallel to the strike of the layers, while the dip slopes are inclined in the direction of dip. The dip slopes become narrower and steeper as the dip of the layers increases, so that the cuesta changes into a strike ridge as the layers approach the vertical. Razorbacks are strike ridges with sharp crests and steep sides, while hogbacks have a broader and more rounded profile.

Structure Contours and Their Use

Although much information can be obtained by studying the stratigraphic column and the structural observations plotted on a geological map, this approach must be supplemented by considering the nature of the outcrop pattern as it is affected by the topography. This allows a much clearer understanding of the geological structure to be gained, based on the use of structure contours.

The boundaries shown on a geological map are simply the surfaces across which the various formations come into contact with one another, wherever these surfaces are seen at the earth's surface. Each boundary represents the geological contact between two adjacent formations. Contours can be used to show the form of such a geological surface in three dimensions, in exactly the same way as they are used to show the form of a topographic surface. They are then known as structure contours, stratum contours or strike lines. They are drawn as lines of equal height above or below some reference level on the surfaces separating geological formations from one another, as shown in Figure 2.8. Since these contacts are not necessarily formed by bedding planes between sedimentary strata, it is better to refer to these lines as structure contours rather than stratum contours.

Structure contours are sometimes known as strike lines because the horizontal trend of the contour lines gives the strike of the contoured surface as it varies from point to point. Likewise, the spacing of the structure contours gives a measure of the dip of the contoured surface as it varies

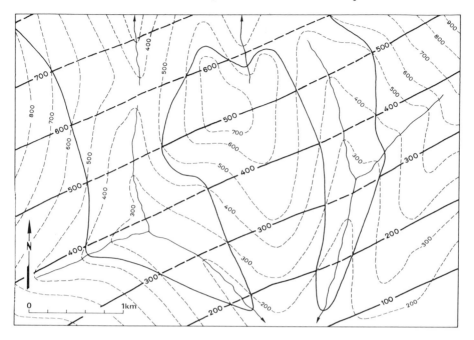

FIG. 2.8. Topographic map showing a series of structure contours representing a geological contact striking ENE or WSW across the area and dipping at a shallow angle towards the SSE. The geological boundary corresponding to this contact is shown by the heavy line. Note that dashed lines are used for the structure contours whenever they lie above the present level of erosion as given by the topographic contours.

from point to point. Structure contours are commonly drawn as equally spaced sets of parallel lines on the problem maps used in practical classes to illustrate the elements of geological map interpretation. Such a contoured surface would have a uniform dip and strike. Such artificiality is virtually never seen in nature, since the contoured surface always shows some variation in its dip and strike as it traced throughout a particular area. This is well shown by the Geological Quadrangle Maps which are published with structure contours by the United States Geological Survey.

Nature of V-shaped Outcrops.

The use of structure contours to show how the outcrop of a geological boundary is affected by the topography can now be considered. Such a boundary represents the line along which a contact between two adjacent formations intersects the earth's surface, where it has been mapped. If structure contours are drawn on this contact, as shown in Figure 2.9, each contour represents a horizontal line lying within this surface at a particular height above Ordnance Datum. Such a line must intersect the earth's sur-

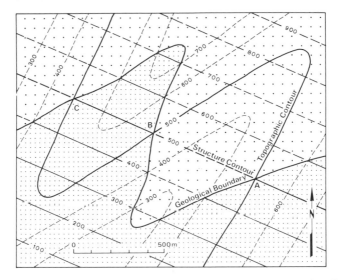

FIG. 2.9. Topographic map showing how the struture contours drawn at 500 m OD on a geological contact intersects the topographic contours at the same height in a series of points A, B and C, through which the geological boundary representing this contact also passes. Note that the same relationship holds for all the other pairs of topograpic and structure contours, drawn at corresponding heights above Ordance Datum.

face at a point where the geological contact also reaches the earth's surface, since it lies within this surface. This means that the structure contour must pass through the same point as the geological boundary representing the contoured surface. However, the topographic contour drawn at the same height as this structure contour must also pass through this point, where the geological boundary representing the contoured surface outcrops at a particular height above Ordnance Datum. Accordingly, three lines must pass through the point of outcrop where they all have the same height above Ordnance Datum: (i) the topographic contour corresponding to the height of this point; (ii) the structure contour drawn at the same height on the contact; and (iii) boundary drawn on the geological map to represent the outcrop of the contact at the surface.

The interpretation of a geological map generally requires that the outcrop pattern be studied in relation to the topography. For example, Figure 2.10 shows the various patterns which can be developed by the outcrop of a geological boundary as it crosses a topographic feature such as a river valley. It represents a contact between two formations with a rather uniform strike form north-east to south-west, except where it is shown as horizontal. The dip varies from the horizontal, through the vertical and back to the horizontal. Structure contours have been constructed for each diagram to show how the outcrop pattern depends on the attitude of the

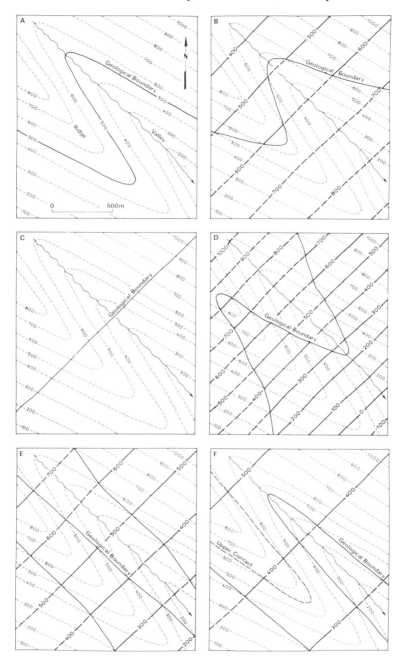

FIG. 2.10. Topographic maps showing how the V-shaped outcrop of a geological boundary varies according to its dip, as shown by the structure contours trending NE-SW across each diagram. See text for a full description.

contact in relation to the gradient of the valley. The following conclusions can be drawn:

(A) If the contact is horizontal, the outcrop of the geological boundary is parallel to the topographic contours. In this case, the outcrop pattern is wholly dependent on the nature of the topography.

(B) If the contact dips upstream, the geological boundary *descends* the valley sides to form a V-shaped outcrop pointing upstream. The V-shape becomes less acute as the angle of dip increases.

(C) If the contact is vertical, the outcrop of the geological boundary is a straight line parallel to the direction of strike. In this case, the outcrop pattern is unaffected by the topography

(D) If the contact dips downstream at a steeper angle than the gradient of the valley, the geological boundary *descends* the valley sides to form a V-shaped outcrop pointing downstream. The V-shape becomes more acute as the angle of dip decreases.

(E) if the contact dips downstream at the same angle as the gradient of the valley, the geological boundary forms two separate outcrops on either side of the valley.

(F) If the contact dips downstream at a shallower angle than the gradient of the valley, the geological boundary *ascends* the valley sides to form a V-shaped outcrop pointing upstream.

Accordingly, if a geological boundary *descends* the valley sides to form a V-shaped outcrop pointing upstream or downstream, the contact dips in the same direction. However, if the geological boundary ascends the valley sides to form a V-shaped outcrop pointing upstream, the contact dips in the opposite direction. The opposite relationships are seen whenever a geological boundary crosses a ridge. It should be noted that the strike of the contact is given by the trend of the geological boundary wherever this is parallel to the topographic contours.

Structure contours should be constructed, at least in imagination, whenever the outcrop pattern is studied to determine the dip of the bedding. A useful rule to remember is that the V-shaped outcrop of a geological boundary across a valley always points towards the area where the contact is present below the topographic surface. Although this direction does not necessarily correspond to the dip of the contact, it is the direction in which the sedimentary rocks overlying the contact are encountered. This means that the V-shape of a geological boundary always points towards the outcrop of the younger formation where it crosses a river valley, provided that the bedding is not inverted, according to the Principle of Superposition.

True and Apparent Dip

The dip of a plane is given in effect by the attitude of a line lying within this plane at right angles to the direction of strike. However, there are any number of other lines inclined at various angles within a given plane. The attitude of these lines gives the apparent dip of the plane in a particular direction, in contrast to the inclination of the line at right angles to the direction of strike, which gives the true dip of the plane. This can be illustrated by means of structure contours.

Figure 2.11 shows a pair of structure contours drawn on a uniformly dipping plane. The true dip can be measured by drawing a line AB at right angles to the contours. The angle α of dip is then given by tan $\alpha = Z/X$

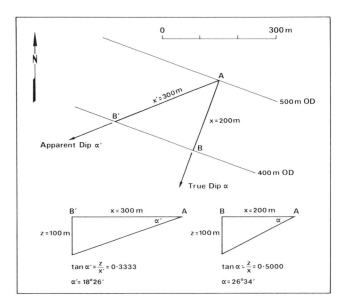

Fig. 2.11. Diagram showing the difference between true and apparent dip. See text for details.

where Z is the contour interval and X is the horizontal spacing of the contours, corresponding to the distance AB. The direction of dip is given by the trend of the line AB, at right angles to the direction of strike, which is given in turn by the trend of the structure contours.

The apparent dip can be found in any other direction by drawing the line AB', parallel to the required direction. The angle α' of apparent dip is then given by tan $\alpha' = Z/X'$ where X' is the horizontal distance between the structure contours in the direction of apparent dip, as given by the trend of the line AB'. It can be seen that the apparent dip varies from zero in the direction of strike to a maximum value where it corresponds to the direction of true dip.

There are two situations where the concept of apparent dip is useful. The first arises in drawing vertical cross-sections which are not parallel to the dip of the sedimentary formations, or any other plane to be shown. Apparent rather than true dips should then be used. Figure 2.12 shows how the apparent dip α' varies as a function of the true dip α and the difference   in direction between the true and apparent dips. The second situation occurs

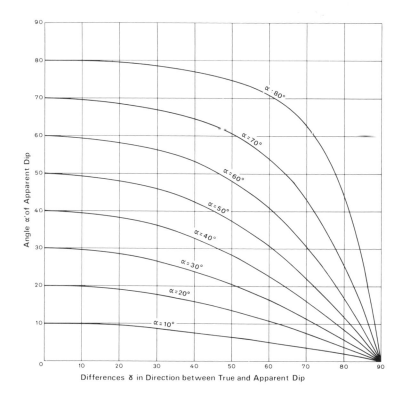

FIG. 2.12. Graph giving the apparent dip α' in terms of the true dip α and the difference γ in direction between the true and apparent dip.

wherever a geological boundary is exposed on a slope. The trend of this geological boundary then corresponds to a direction of apparent dip, while its inclination gives the angle of apparent dip in this direction. This angle can be found simply by measuring the distance X' along the geological boundary between the topographic contours. It is important to realise that a geological boundary always descends a valley side in a direction of apparent dip at an angle which increases as this direction becomes closer to the direction of true dip. Conversely, the strike of the contact under consideration is given by the trend of a geological boundary wherever it runs parallel to the topographic contours.

Exercises Using Structure Contours

A number of different exercises can be undertaken using structure contours as an aid in visualising how the topography affects the outcrop pattern developed by the dip of sedimentary rocks. They illustrate the methods used to find the dip of a geological contact from a study of the outcrop pattern, taking the topography into account. Alternatively, they can be used to determine the effect of topography on the outcrop of a geological boundary, knowing the dip of the contact in question. These exercises are usually done on what are known as problem maps, which are commonly constructed in such a way that the structure contours are drawn as straight and parallel lines at a constant spacing from one another. It should be emphasised that this is only done to provide a unique solution to the problem. Many of these maps can be adapted to show a more realistic pattern of structure contours, using published maps as a basis. The Geological Quadrangle Maps published by the United States Geological Survey are particularly good for this purpose.

Drawing in Structure Contours.

Any contact between two formations can be contoured by considering how the topography affects the outcrop of the geological boundary corresponding to this contact. This is done simply by taking a topographic contour of a particular height above Ordnance Datum and marking off all those points where it intersects the geological boundary. The structure contour for the same height can then be drawn as a smooth curve through all these points. It is best to start with the topographic contour which gives the widest spacing for the points of intersection. This procedure is then repeated for each topographic contour in turn until all the structure contours possible have been constructed for the contact in question. It is commonly found that these contours are rather uniform in trend and spacing, so that the geological boundary corresponds to a contact with a rather uniform dip and strike. Further contours can then be constructed by extrapolation, to cover the whole map. This construction is obviously much simpler if the contact in question has a uniform dip and strike. The structure contours would then be a set of straight and parallel lines, lying at a constant distance from one another. The dip and strike of the contact can be determined in either case, using the methods already described.

Drawing in a Geological Boundary.

The outcrop of a geological boundary can be constructed from a topographic map, given the structure contours for the contact in question. This is possible because the boundary must pass through all those points where equivalent pairs of structure and topographic contours, having the same height above Ordanance Datum, intersect one another. It is then possible to determine the outcrop of a geological boundary given the topographic and structure contours, simply by drawing a line through all those points where the topographic and structure contours have the same height above Ordnance Datum.

It is usually best to start drawing in the geological boundary on the side of a valley, where it would trend in a particular direction according to its apparent dip. Note that the geological boundary changes slightly in direction wherever the spacing of the topographic contours changes. The boundary should be drawn as a smooth curve to accommodate such changes in direction. If the topographic contours are not straight, the geological boundary would be drawn as a curved line which gradually cuts across the topographic contours at a uniform angle.

The geological boundary can then be followed across the map changing direction wherever it crosses intervening valleys and ridges. It should be drawn so that the geological boundary never crosses a topographic contour without intersecting a structure contour at the same height, and vice versa. Where a topographic contour touches a structure contour tangentially, the geological boundary should be drawn so that it also touches the structure contour tangentially. This means that the boundary would be curved in such a way that it did not cut across the structure contour.

It is important to realise that the geological boundary drawn in this way for a particular contact can have more than one line of outcrop. Such a situation would occur wherever a geological boundary outcrops on the floor of a valley or at the top of a hill, away from its main outcrop, so forming an inlier or an outlier. It is then found that the geological boundary, located in such a position by the intersection of corresponding topographic and structure contours, could not be joined to the main line of outcrop without transgressing the rules already given for its construction. This means that it should then be drawn as a separate line of outcrop, encircling the valley floor or the hill top, as the particular case requires. It is important in undertaking the construction of a geological boundary to check that each structure contour intersects a topographic contour of the same height in a series of points through which a geological boundary has been drawn. Otherwise, it is possible to miss out the geological boundary wherever it forms a separate line of outcrop.

Determining Stratigraphic Thicknesses.

Only a single horizon has been considered so far. However, the outcrop pattern developed by even the simplest of geological maps is generally formed by a single formation with its upper and lower contacts parallel to one another. The structure contours drawn on these contacts would only differ from one another in their relative position, provided that they had the same dip and strike as shown in Figure 2.13.

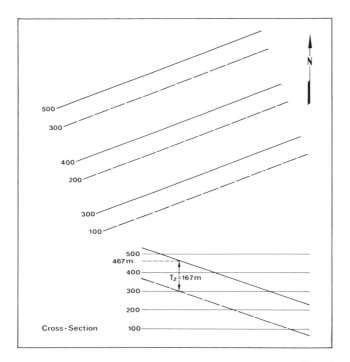

Fig. 2.13. Plan showing structure contours drawn at a 100 m interval on the upper contact (solid lines) and the lower contact (dashed lines) of a parallel-sided formation. Cross-section shows how the vertical thickness T_z can be found from the difference in height of the two contacts at any one point.

The vertical distance between the two contacts can be found by taking a structure contour drawn at a particular height on one contact. The height of the other contact corresponding to the position of this structure contour can then be found by interpolation, using the structure contours drawn on the other contact. The difference in height between the two sets of structure contours then gives the vertical thickness T_z of the formation. It should be noted that the two sets of structure contours would coincide with one another, while differing in height, wherever the vertical distance between the two contacts is a whole-number multiple of the contour interval. Once the vertical thickness of the stratigraphic formation has been found, this can be converted according to its angle α of dip to give the true thickness T_n of the stratigraphic formation, using the relationship $T_n = T_z \cos \alpha$.

This way of measuring the true thickness of a stratigraphic formation can be simplified. Suppose that a set of structure contours has been drawn on the lower contact of the formation. By following a particular contour across the map, it will be found to intersect the geological boundary representing the upper contact of the formation. The height of this point can then be found using the topographic contours. The vertical thickness T_z is given by the difference in height between this point and the structure contour drawn through this point on the lower contact of the formaiton. The true thickness T_n can then be determined, as previously. This method relies on measuring vertical distances between successive contacts in the direction of their strike, as given by the trend of the structure contours.

Completing a Geological Map.

A common problem requires that the structure contours, drawn in using a geological boundary shown on a problem map, are then used to draw in another geological boundary. This is known as completing the geological map. It is usually assumed that the two contacts in question are separated vertically from one another by a distance which is equal to the contour interval or its multiple. The two sets of structure contours then coincide with one another, although they differ in height. The actual difference in height may be provided by the log of a borehole, which would give the vertical distance between the two horizons. Alternatively, it can be given by specifying a point where the second contact is known to outcrop. The geological boundary to be constructed would then pass through this point.

A set of structure contours are first drawn, generally as a series of straight and parallel lines on the boundary shown by the geological map. If the outcrop of the other boundary is given, it will generally be found that one contour passes through a topographic contour at the point where this boundary is known to outcrop. This contour can then be taken as a structure contour drawn on this boundary at the same height as the topographic contour in question. The difference in height between these two structure contours gives the vertical thickness of the intervening formation. This may be given directly by the borehole information. The heights of all the other structure contours can then be altered by a similar amount, so allowing the geological boundary to be drawn in using these contours. This completes the geological map, as required by the problem.

Three-point Problems.

Structure contours can also be constructed if the height of a contact is known at three points. This is done by drawing straight lines between the three points as shown in Figure 2.14. Each line is then divided into segments by interpolating a series of points corresponding to the required heights of the structure contours. Each contour is then drawn as a straight line passing through the points of equal height. This construction can be simplified if one of the original points has a height corresponding to a structure contour.

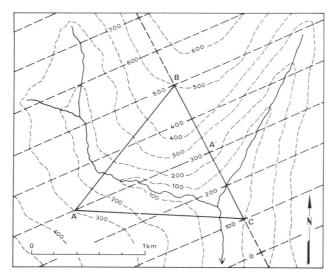

Fig. 2.14. Topographic map showing how structure contours can be constructed for a geological boundary which is known to pass through the points A, B and C.

It assumes that the contact is a planar surface since the structure contours can only be drawn as straight lines, equally spaced and parallel to one another. The geological boundary can then be drawn in using these structure contours, as previously described.

Construction of Vertical Cross-sections.

The exercises based on problem maps often require that a vertical cross-section should be drawn to illustrate the geological structure, using the structure contours which have been constructed from the outcrop pattern. Such a cross-section should be drawn along a line at right angles to the strike, so that it shows the true dip of the stratigraphic formations. Once the line of cross-section has been chosen, two vertical lines are drawn on a piece of graph paper, corresponding to the ends of the cross-section, as shown in Figure 2.15. A vertical scale is marked off along one of these lines, equal to the horizontal scale of the map, so that the cross-section will be drawn to scale. This scale should be divided into intervals corresponding to the contour heights shown on the map. A horizontal line is then drawn between the two vertical lines to give the height of the Ordnance Datum.

The points where the topographic contours intersect the line of cross-section are then marked off along a strip of plain paper. The height of each point should be noted. The strip is then used to transfer each point to its correct position on the cross-section, as shown for the 800-m contour in Figure 2.15. Each point transferred in this way should be clearly marked on

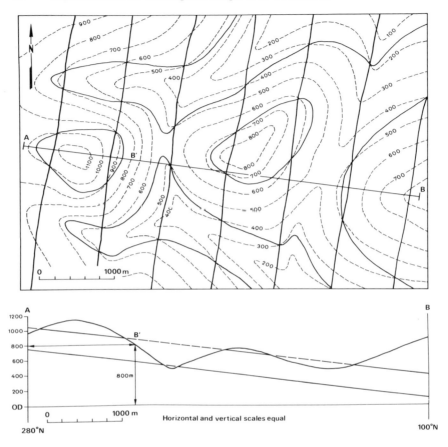

Fig. 2.15. *Upper diagram:* geological map with two contacts separated from one another by a vertical distance of 300 m. *Lower diagram:* vertical cross-section along the line AB. Note how the position of the point B′ where the 800 contour crosses the line of cross-section is located by measuring off the distance AB′ along a horizontal line at a height corresponding to 800 m OD on the cross-section. This is repeated for all the other points in which the topographic contours cross the line of the cross-section, so allowing the topography to be drawn in.

the cross-section. A line can then be drawn through all these points to give a vertical cross-section through the topography as required.

The structure contours can be transferred in exactly the same way from the geological map to give a vertical cross-section through the geological structure defined by these contours. The geological contacts so drawn are usually shown as dashed lines wherever they have been extrapolated above the topographic surface. This manner of representation can be extended to structure contours. They are usually shown as solid lines wherever these contours are drawn on a geological contact which lies below the topographic surface but as dashed lines wherever this contact has been lost by erosion from above the topographic surface. Figure 2.8 shows that the outcrop of a geological contact forms a boundary separating these areas from one another.

The vertical cross-section constructed in this way should always be drawn as far above the height of the topographic surface, and as far below the level of the Ordnance Datum, as the evidence provided by the geological map will allow. There is no justification for only showing the geological structure from the topographic surface downwards to the level of the Ordnance Datum. Likewise, vertical cross-sections should be drawn with the horizontal and vertical scales equal, so that there is no vertical exaggeration to distort the true form of the geological structures. This is particularly important wherever the cross-section shows folded and faulted rocks, since changing the vertical scale in this way causes the dips to become exaggerated. However, it is sometimes necessary to introduce a moderate degree of vertical exaggeration into cross sections showing flat-lying formations in areas of low relief particularly if the formations are relatively thin horizons. Otherwise, it would not be possible to show the geological structure in any detail at all.

CHAPTER 3

Folds and Folding

Geometry of Folded Surfaces

FOLDS are developed in rocks wherever a pre-existing surface becomes distorted into a wavy or zig-zag form, as shown in Figure 3.1. Folding may affect any type of rock, provided that there is some structure already present to register its effects. However, folds are best displayed by layered

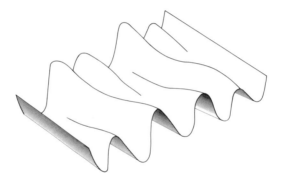

FIG. 3.1. Diagram of a folded surface in three dimensions, typical of cylindroidal folding.

sequences of sedimentary rocks, where it is the bedding planes separating the individual beds from one another which become folded. Unless these beds were deposited with an initial dip, it can be assumed that the folded surfaces were originally horizontal, thus allowing some measure of the movements involved in the folding to be determined. This is not possible wherever surfaces other than bedding are folded, since their original attitude would not be known with any certainty. Moreover, the example set by the folding of sedimentary rocks should not be taken to suggest that folding can only affect planar surfaces, which were originally parallel to one another, even although this usually appears to be the case.

The configuration of a folded surface is best described in terms of its curvature. This can be measured in two dimensions by considering how rapidly a curve deviates from its tangent as it is traced away from a particular point

(See Fig. 3.2A). If a surface is folded in three dimensions, its curvature can vary as it is measured in any direction passing through a particular point. In general, these measurements at any one point on a folded surface vary from a maximum to a minimum. The directions of maximum and minimum curvature at such a point lie at right angles to one another, as shown in Figure 3.2B. These directions, and the amount of curvature parallel to each direction, vary from point to point on the folded surface, so defining its form, as shown in Figure 3.1. The wave-like undulations developed in such a surface reflect how the sense of curvature changes repeatedly from concave to convex, and back again, along these directions of maximum and minimum curvature.

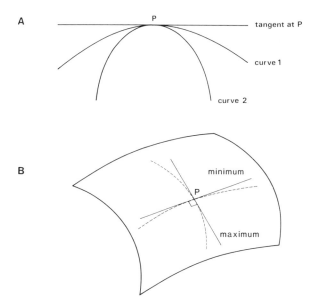

Fɪɢ. 3.2. A: Diagram showing how two curves deviate from the tangent at a point P as a measure of their curvature. Curve 2 has the greater curvature. B: Diagram showing directions of maximum and minimum curvature in three dimensions.

Cylindrical Folds.

The simplest form of a folded surface is developed if the directions of maximum curvature are normal to a constant direction of zero curvature. Such a surface can be generated as a series of cylindrical segments by a straight line moving at right angles to itself along a curved path (See Fig.3.3A). The folds so formed are known as cylindrical folds. The straight line generating the folded surface is termed the fold axis. It corresponds to the direction of zero curvature in the folded surface. The orientation of the fold axis in space defines the plunge of the fold.

There is only one cross-section through a cyindrical fold which provides a true picture of its shape. This cross-section, drawn in the plane at right angles to the fold axis, is known as the fold profile (see Fig. 3.3B). It should be noted that the profile is simply defined by the curved path along which the line generating a folded surface moves at right angles to itself. Since this path could be traced out by any point on the line generating the folded surface, the profile must remain the same in the direction of the fold axis. This is an important feature of cylindrical folds.

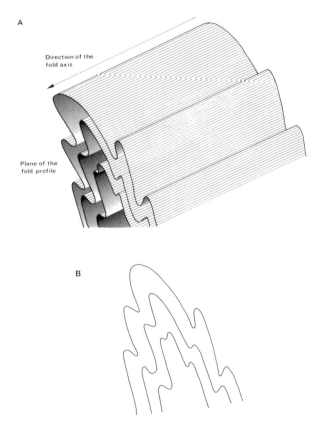

A

Direction of the
fold axis

Plane of the
fold profile

B

FIG. 3.3. A: Diagram of a cylindrical fold, showing the direction of the fold axis. Note that the folded surfaces can be generated by a straight line moving at right angles to itself, as shown by the ruled lines drawn on the upper surface, parallel to the direction of the fold axis. B: The fold profile drawn in a plane at right angles to the fold axis.

Cylindroidal Folds.

The regular form of a surface affected by cylindrical folding becomes distorted as the amount of curvature normal to the direction of maximum curvature increases from zero. Accordingly, a series of irregular undulations would be developed in the folded surface, giving rise to non-cylindrical folds. However, if the amount of curvature is much less in the direction of minimum curvature at right angles to the direction of maximum curvature, the surface will still retain a form which is approximately cylindrical (see Fig. 3.1). Indeed, such surfaces can often be divided into a series of cylindrical segments, which only differ slightly in orientation from one another. The fold axes describing the orientation of these segments differ in attitude from one another. However, they can be taken to define the direction of the mean fold axis. This is defined as the closest approximation to a straight line which would generate the folded surface if it was moved at right angles to itself along a curved path. The individual folds developed in such a surface tend to be parallel to one another, even although they can rarely be traced for any distance in the direction of the mean fold axis. Such folds are best termed cylindroidal folds according to the meaning given to this suffix by the *Oxford English Dictionary*, despite the fact that this adjective has previously been used in the literature as a synonym for "cylindrical".

The concept of the fold profile as the true cross-section through a folded surface can only be applied strictly to cylindrical folds. However, profiles can be drawn across cylindroidal folds in a plane at right angles to the mean fold axis. This gives the closest approximation to a true cross-section through the folded surface. Cylindroidal folds differ from cylindrical folds in that they do not maintain the same fold profile in the direction of the mean fold axis. Indeed, how far cylindroidal folds depart from the ideal geometry of cylindrical folds can be judged according to how their fold profile changes in this direction.

Non-cylindroidal Folds.

Such folds become increasingly non-cylindroidal as the curvature in the direction of minimum curvature becomes greater. This leads to the wave-like undulations in the folded surface being replaced by more irregular structures, which generally occur in the form of domes and basins, as shown in Figure 3.4. The curvature varies in a complex way across such a surface. However, individual domes and basins define areas where the curvature is more-or-less equal in all directions, while the saddles separating these domes and basins from one another form areas where the curvature has the opposite sense in the two directions at right angles to one another.

Even although the folding is non-cylindroidal, there may still be regions of the folded surface which have approximately a cylindroidal form. However, other parts of the folded surface may be conical in form. Conical folds are analogous to cylindrical folds in that the curvature is zero at right angles to the directions of maximum curvature in the surface. However, in a conical fold, these directions of zero curvature form a series of converging

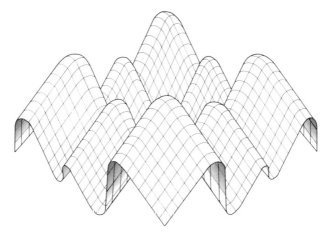

FIG. 3.4. Diagram of non-cylindroidal folding, showing the development of domes and basins in a folded surface.

lines, so that they are not parallel to one another. Although conical folds in the strict sense appear to be very rare, the terminations of cylindroidal folds may have approximately a conical form.

Description and Classification of Folds

Since folds are perhaps the most varied of geological structures, they require an exact and comprehensive terminology for their description and classification. However, many terms introduced into the literature were not defined with enough precision to be used without ambiguity. Some definitions were founded on too simple a view of the complex geometry shown by folded surfaces, while others had genetic connotations that are inappropriate to any scheme which attempts to describe and classify folds in an objective manner. The terms based on such definitions have a tendency to become obsolete with the advance of geological knowledge. Moreover, some terms were not used in the same sense by different authors, while others have changed in meaning from their original definition. Even so, a concensus has generally been reached among structural geologists concerning the essential terminology to be used in describing and classifying folds. Although it is based on concepts which arise as a result of cylindrical folding, it may be applied with little modification to cylindroidal folds.

Plunge and Pitch.

The terminology to be described in this section recognises various planar and linear elements in a fold. The attitude of a plane may simply be recorded in terms of its dip and strike, as defined in Chapter 2. Likewise, the

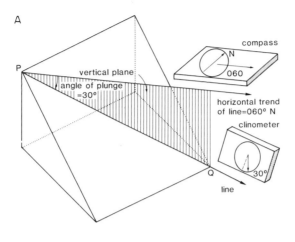

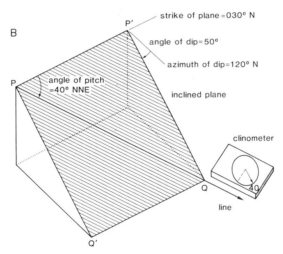

FIG. 3.5. A: Plunge of the line PQ within a vertical plane, showing how to measure its attitude in the field using a compass and clinometer. B: Pitch of the line PQ within the inclined plane PP′QQ′ showing how to measure its attitude in the field with clinometer. clinometer.

attitude of a line in space may be measured by either its plunge within a vertical plane or its pitch (rake) within an inclined plane, as shown in Figure 3.5. Both measurements are made *downwards* from the horizontal.

The plunge of a line is measured with reference to the vertical plane which contains the line (see Fig. 3.5A). There is only one such plane, so that its attitude need not be specified. The angle of plunge is then given by the angle between the line and the horizontal, as measured within this vertical plane, while the direction of plunge is given by the horizontal trend of the line within this plane, as measured by its azimuth from true north. The latter measurement corresponds to the strike of the vertical plane in the down-plunge direction of the line. For example, a line may be said to plunge at 30° towards 060°N.

The pitch of a line is measured with reference to an inclined plane which contains the line (see Fig. 3.5B). The attitude of this plane must be specified in terms of its dip and strike. The angle of pitch is then given by the angle between the line and the horizontal, as measured within this plane, while the direction of pitch corresponds to the strike of this plane from which the angle of pitch was measured, as given simply by a compass direction. For example, a line may be said to pitch at 40°NNE within a plane dipping at 50° twoards 120°N. Note that the direction of pitch must be specified to distinguish between the two lines lying at the same angle from the horizontal within an inclined plane. thus, there is also a line P′Q′ pitching at 40°SSW within a plane dipping at 50° towards 120°N, which can only be distinguished from the line previously mentioned by specifying the direction of pitch as SSW rather than NNE.

The plunge and pitch of a line can be measured in the field, using a compass and clinometer. The plunge is measured by holding the clinometer vertically so that its long edge is parallel to the line. The angle of plunge can then be measured from the horizontal. A compass is then held against the clinometer, and the direction of plunge measured. Likewise, the pitch of a line within a plane can be measured by placing the clinometer on the plane so that its long edge is parallel to the line. The angle of pitch is then measured from the horizontal within this plane. The dip and strike of this plane must also be measured, as described in Chapter 2. This allows the direction of pitch to be noted as corresponding to the direction of strike from which the angle of pitch was measured.

Plunge of a Fold.

The hinge of a fold is defined as the line joining the points of maximum curvature in the folded surface, as shown in Figure 3.6. Hinge line and fold hinge are equivalent terms. So defined, the hinge may be regarded as the line around which the surface is folded. This line corresponds, at least approximately, to a direction of minimum curvature in the folded surface.

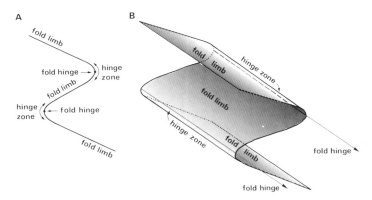

FIG. 3.6. A: Definition of the fold hinge as the point of maximum curvature within the fold profile. The hinge zones separate adjacent fold limbs. B: A cylindrical fold in three dimensions, showing the rectilinear nature of the fold hinges.

Cylindrical folds have straight hinges which, being parallel to the fold axis, are parallel to one another. Indeed, "fold axis" has been used in the past as a synonym for "fold hinge", even although a clear distinction is now drawn between the two terms. According to this distinction, a fold axis refers to the direction of a line which can be taken to generate a folded surface, so that it is essentially a reference axis in space, while the fold hinge is a discrete line which can be identified according to its position in the folded surface.

Cylindroidal folds differ from cylindrical folds in that they have curved hinges, as shown in Fig. 3.7, which only trend approximately in the same direction as the mean fold axis. Indeed, how far cylindroidal folds depart from the ideal geometry of a cylindrical fold can be judged by the amount of curvature shown by the hinges of individual folds within a particular region. This can be measured by the dispersion of these fold hinges from a common attitude, which would correspond to the direction of the mean fold axis. The highest and lowest points on a folded surface in the plane of the

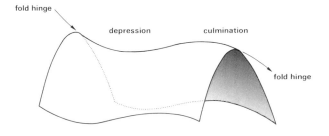

FIG. 3.7. Cylindroidal fold with a curved hinge, showing the development of culminations and depressions.

fold profile are termed the crest and trough, respectively (see Fig. 3.8). If the folding is non-cylindrical, culminations and depressions are commonly developed along the crests and troughs of individual folds (see Fig. 3.7).

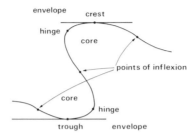

FIG. 3.8. Definition of crest and trough in a fold profile.

There is a standard classification used to describe the attitude of a fold in terms of the plunge shown by its hinge, as follows:

Horizontal	0° to 10°
Gently Plunging	10° to 30°
Moderately Plunging	30° to 60°
Steeply Plunging	60° to 80°
Vertical	80° to 90°

Accordingly, horizontal folds have hinges plunging at less than 10°, plunging folds have hinges plunging between 10° and 80°, and vertical folds have hinges plunging at more than 80°.

Inclination of a Fold.

The surface passing through successive hinges in a series of folded surfaces is known as the axial plane or the axial surface (see Fig. 3.9). Although the latter term should perhaps be used if this surface is not a plane, it is common practice to use the former term even for curved surfaces. This has an advantage since it allows "axial-planar" to be coined as an adjective, which would otherwise be lacking. The axial plane of a fold, by definition, intersects each of the folded surfaces in a line which coincides with the corresponding fold hinge, as shown in Figure 3.9. It is important to realise that this definition does not imply in any way that the axial plane of a fold bisects the angle between its fold limbs.

Folds commonly have axial planes which are virtually parallel to one another, as shown in Figure 3.10A. However, it is possible for adjacent folds to have axial planes which are inclined towards one another at a high angle, giving rise to conjugate folds or box folds (see Fig. 3.10B). The

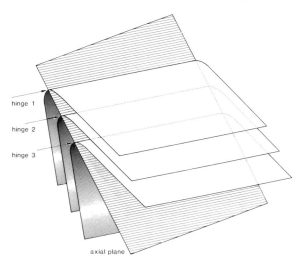

Fɪɢ. 3.9. Axial plane of a fold passing through successive hinges in a series of folded surfaces.

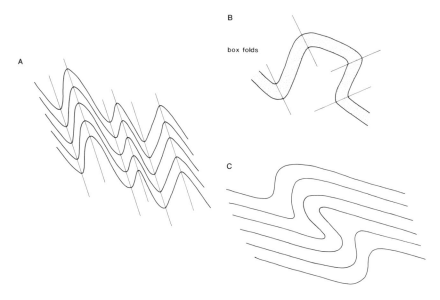

Fɪɢ. 3.10. A: Folds with parallel axial-planes. B: Box folds with conjugate sets of axial planes. C: Disharmonic folds.

distinction between the two types is that conjugate folds generally occur on a smaller scale than box folds. It is also possible for the folded surfaces to change rapidly in profile along the axial plane of a fold at right angles to the fold axis, giving rise to disharmonic folds, as shown in Figure 3.10C.

There is a standard classification used to describe the attitude of a fold in terms of the dip shown by its axial plane, as follows:

Upright	90° to 80°
Steeply Inclined	80° to 60°
Moderately Inclined	60° to 30°
Gently Inclined	30° to 10°
Recumbent	10° to 0°

Accordingly, upright folds have axial planes dipping at more than 80°, inclined folds have axial planes dipping between 80° and 10°, and recumbent folds have axial planes dipping at less than 10°.

Folds vary in inclination according to the dip of their axial planes. However, it is particularly convenient to define the inclination of a fold as the direction opposite to the dip of its axial plane, as suggested by the everyday meaning of this word. Thus, a fold would be inclined towards the northwest if its axial plane dipped towards the south-east.

General Classification of Fold Attitude.

Any fold can be considered to vary in attitude according to the plunge of its fold hinge and the dip of its axial plane. However, since the fold hinge lies within the axial plane, vertical folds must have upright axial planes while recumbent folds must have horizontal fold hinges, even although the reverse is not the case. Apart from these exceptions, all folds require two sets of terms to describe their attitude, giving rise to the general classification of fold attitude as shown in Figure 3.11. For example, a fold can be described as gently plunging but steeply inclined.

Since the plunge of the fold hinge cannot exceed the dip of the axial plane, certain terms cannot be combined with one another. For example, folds cannot be described as steeply plunging but gently inclined. Moreover, as the plunge of the fold hinge approaches the dip of the axial plane, the pitch of the fold hinge within the axial plane increases to 90°. This means that the fold would plunge down the dip of its axial plane. Such folds are termed reclined folds. They are intermediate in attitude between vertical and recumbent folds. The variety of possible orientations is shown in Figure 3.12, drawn to correspond directly with the classification of Figure 3.11.

The trend of a fold can be used to describe the plunge of the fold hinge. However, it may also refer to the strike of the axial plane. The latter connotation arises from the fact that the axial plane outcrops along a line which is known as the axial trace of the fold. The trend of this line gives the strike of the axial plane wherever the topography is horizontal. It should be clearly understood that the plunge of the fold hinge is only parallel to the strike of the axial plane if the fold hinge is horizontal or if the axial plane is

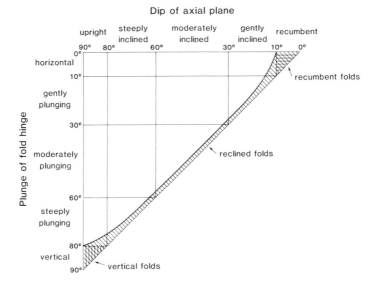

FIG. 3.11. Classification of fold attitude (after Fleuty). Note that reclined folds have been defined as closing within 10° of the horizontal within their axial planes.

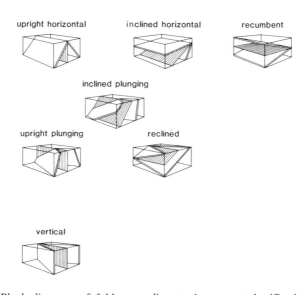

FIG. 3.12. Block diagrams of folds according to the present classification of fold attitude.

vertical. Otherwise, it is generally the case that these directions are oblique to one another, wherever the fold hinge plunges at an angle within an inclined axial plane. Reclined folds represent an extreme case, where the fold hinge plunges at right angles to the strike of the axial-plane.

This means that simply describing the trend of a fold without reference to the plunge of the fold hinge or the strike of the axial plane can only lead to confusion. It is therefore emphasised that considerable care must be taken in order to describe the orientation of a fold unambiguously. This is best done by providing a full description of the orientation shown by the various elements which define the fold, so that a complete picture of the fold can be constructed.

Tightness and Angularity.

The region near the hinge where the folded surface has an appreciable curvature is termed the hinge zone. The hinge zones of adjacent folds in a surface are separated from one another by the fold limbs, where the curvature of the folded surface is much less. There is generally a line of inflexion on each fold limb where the curvature changes from concave to convex, or vice versa, as already shown in Figure 3.8. The surface folded around the hinge zone separates the core of the fold on its concave side from the envelope of the fold on its convex side. The hinge zone is sometimes known as the nose of the fold.

The tightness of a fold can be described by the angle between the fold limbs, as measured between the lines of inflexion in the plane of the fold profile. This gives the inter-limb angle of the fold, as shown in Figure 3.13A. The following scale is used to define the tightness of a fold in terms of its inter-limb angle:

Gentle	180° to	120°
Open	120° to	70°
Close	70° to	30°
Tight	30° to	10°
Isoclinal	10° to	0°

In certain types of fold, the inter-limb angle is negative in that the folded surface passes through an angle greater than 180° as it is traced around the fold hinge. Such folds are known as elasticas, as shown in Figure 3.13B.

The angularity of a fold depends on the relative width of the hinge zone. Most folds have rounded hinges across which there is a gradual change in the attitude of the folded surface. However, there is a distinctive group of angular folds with straight limbs and narrow hinges. Such structures are generally known as kink folds or chevron folds, according to their size. Convention generally dictates that kink folds are developed on the scale of a

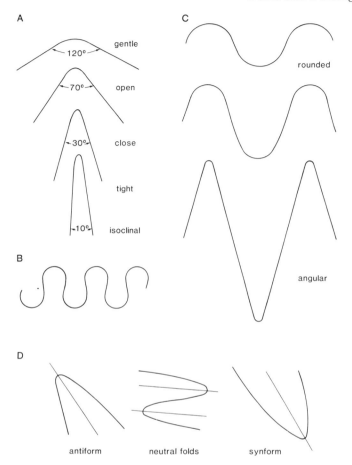

FIG. 3.13. A: Diagram showing the values of interlimb angle which serve to define the tightness of a fold. B: Elasticas in a folded surface. C: Diagram showing how fold hinges can vary from rounded to angular. D: Definition of folds as antiforms, synforms or neutral folds.

few centimetres or less, whereas chevron folds are considerably larger. A kink band is formed by a pair of kink folds maintaining the same profile through all the surfaces along a closely spaced and parallel set of axial planes. A strain band rather than a kink band is developed if the folds occurring close together in this way have rounded hinges. Kink bands in particular tend to form conjugate structures at a high angle to one another.

Direction of Fold Closure.

Cylindroidal folds can be named according to the direction in which the fold limbs converge on one another at the fold hinge. This direction defines the closure of the fold, as measured within the axial plane at right angles to the fold hinge. The direction of fold closure therefore lies within the plane of the fold profile.

Although folds have been described in the past as "anticlines" or "synclines" according to the direction of fold closure, it is now accepted that these terms should only be used in a stratigraphic sense to indicate whether older or younger rocks occur in the core of a fold. These terms have therefore been replaced by "antiform" and "synform". Cylindroidal folds closing upwards in the plane of the fold profile are termed antiforms, whereas those closing downwards in the same plane are known as synforms, as shown in Figure 3.13D. The limbs of an antiform dip away from the axial plane, whereas the limbs of a synform dip towards the axial plane, even if they are inclined in the same direction. However, if fold limbs are inclined in the same direction, one or other has been overturned. Such a fold is known as an overfold if it is formed by an antiform, whereas it might be termed an underfold if it is formed by a synform. Such folds can, however, be more simply described as overturned antiforms or synforms as the case may be.

A monocline is formed wherever a zone of steeply dipping strata affects a flat-lying sequence of sedimentary rocks. Alternatively a structural terrace is produced by a local flattening of the strata in an area where otherwise the bedding is more steeply inclined in a single direction.

Some folds close sidewards, rather than upwards or downwards, in the plane of the fold profile, so that they are neither antiforms nor synforms. Such folds are known as neutral folds. They form a separate class, consisting of all those folds which close in a horizontal direction within the plane of the fold profile, as shown in Figure 3.13D. Recumbent, reclined and vertical (plunging) folds are all examples of neutral folds, closing sidewards within their axial planes. Indeed, a reclined fold can be strictly defined as an inclined fold which closes within 10° of the horizontal.

The names given to non-cylindroidal folds have a similar connotation. Thus, a structural dome corresponds to an antiform in that it closes upwards, whereas a structural basin corresponds to a synform in that it closes downwards. Intermediate types of fold are represented by brachy-anticlines and brachy-synclines, which should more properly be termed brachy-antiforms and brachy-synforms if only their structural form is under consideration. It is a characteristic feature of such structures that their hinges pass through the horizontal to plunge in opposite directions. They are commonly developed wherever antiforms and synforms show marked culminations and depressions along their length. All these structures belong to the general class of periclinal folds, so called because the dips are inclined in all directions about a central area.

Anticlines and Synclines.

Most folds affect stratigraphic sequences of sedimentary or volcanic rocks. It has already been mentioned that such folds are classified as anticlines or synclines according to the relative age of the rocks affected by the folding. Thus, an anticline is defined as a fold with a core of older rocks and an envelope of younger rocks, in distinction to a syncline which is defined as a fold with a core of younger rocks and an envelope of older rocks, as shown in Figure 3.14. This means that folds can only be recognised as anticlines or synclines if the stratigraphic order to the rocks can be determined.

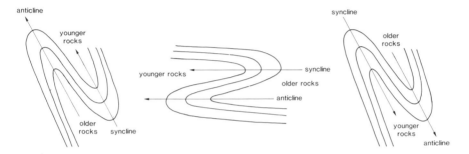

FIG. 3.14 Definition of folds as anticlines and synclines, according to the stratigraphic order of the folded rocks.

There is often no reason to assume *a priori* that the rocks have not been inverted by the folding, unless the folds are rather open structures, so that the Principle of Superposition cannot always be applied. However, it may be possible to recognise the stratigraphic sequence in comparison with an area where the stratigraphic order is known. Alternatively zone fossils may be used to establish the stratigraphic order of the formations affected by the folding, or sedimentary structures may be used to determine the way-up of the individual beds within these formations. The use of sedimentary structures is particularly important in this context, simply because it can be applied to rocks in which zone fossils are either absent or, if present, cannot be recognised. Thus, zone fossils are only found in Phanerozoic rocks, while they are more easily obscured or destroyed by deformation and metamorphism than sedimentary structures.

Facing of Folds.

The way-up and younging of sedimentary rocks is obviously affected by folding. For example, the beds in a fold will young in different directions, as they are traced around the fold hinge. However, if the plane of the fold profile is considered, all these beds become younger in the same direction, parallel to the axial plane of the fold. The fold is therefore described as facing in that direction, normal to the fold hinge but parallel to the axial plane, in which the beds become younger, as shown in Figure 3.15. This is an important concept since a single set of folds will generally face in the same direction, even although the beds affected by the folding young in different directions. This introduces and element of simplicity into the complexity otherwise shown by folded rocks.

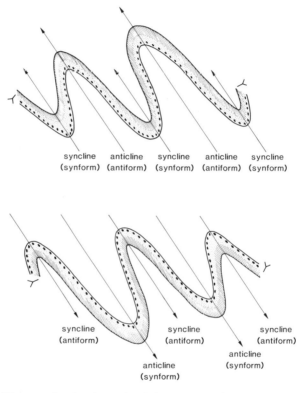

syncline   anticline   syncline   anticline   syncline
(synform)   (antiform)   (synform)   (antiform)   (synform)

syncline   syncline   syncline
(antiform)   (antiform)   (antiform)

                                          anticline
                                          (synform)

          anticline
          (synform)

FIG. 3.15. Diagram showing how the facing of a set of folds depends on the stratigraphic order, as indicated by a graded bed. The arrows give the direction of structural facing, while the ⅄-shaped symbols give the direction of stratigraphic younging.

It has already been mentioned that folds can be classified as antiforms, synforms or neutral folds according to the direction of fold closure, independently of whether they can be recognised as anticlines or synclines on the basis of stratigraphic relationships. However, the direction of facing shown by a fold depends on both factors. Thus, antiforms face upwards if they are anticlines but downwards if they are synclines, whereas synforms face upwards if they are synclines but downwards if they are anticlines.

Recumbent, reclined and vertical folds face sideward in one direction or another, according to whether these neutral folds are anticlines or synclines.

It is commonly the case that folding affects straigraphic sequences which have not been inverted, at least overall. Since the folds would then face upwards, anticlines and synclines are represented by antiforms and synforms, respectively. However, it cannot always be assumed that this is the case as downward-facing folds can be developed if the structural relationships of the folded rocks are more complex. This means that stratigraphic evidence must always be used to establish a fold as an anticline or a syncline, while structural evidence is used to find whether it is an antiform, a synform, or a neutral fold. The facing of the fold can then be determined.

It should be pointed out that a clear distinction between younging and facing, as these terms are used in this book, has not been made in the literature. Thus, facing has been used as a term to describe the younging of stratigraphic sequences as well as the orientation of the folds which affect such sequences. However, it is possible for a folded sequence to young overall in one direction, while the folds affecting such a sequence face in the opposite direction. This can arise simply because the younging of a stratigraphic sequence is described with reference to the horizontal, while the facing of the folds affecting such a sequence is referred to an oblique direction, parallel to their axial planes. Obviously, confusion or uncertainty can result if facing is used as a synonym for younging in such a way that this term has two distinct meanings. It is therefore argued that a clear distinction should be made between younging and facing. This may be emphasised in diagrams by using a λ-shaped symbol pointing in the direction of younging at right angles to the bedding, while an arrow is used to show the direction of structural facing parallel to the axial planes of the fold, in the manner of Figure 3.15.

Morphology of Folded Layers

Most folds affect layers which, differing in lithology from one another, are separated by discrete surfaces. For example, the bedding of sedimentary rocks is formed by discrete bedding planes which separate the different beds from one another. An important feature of fold morphology concerns the way in which these layers vary in thickness as a result of the folding. This depends on how the surfaces forming the boundaries of these layers are folded in relation to one another. A general classification of fold morphology, which is based on such considerations, will be described in this section. However, two special cases can be considered first, simply because most cylindroidal folds closely resemble one or other of these fold types (see Fig. 3.16).

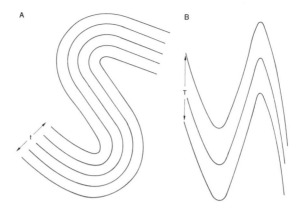

Fɪɢ. 3.16. A: Parallel fold showing how the thickness *t* of the layers normal to the bedding is approximately constant. B: Similar fold showing how the thickness *T* of the layers parallel to the axial planes is approximately constant.

Parallel and Similar Folds.

Parallel folds are defined as folds in which thickness *t* of the folded layers, as measured normal to the folded surfaces, remains the same, as shown in Figure 3.16A. They are sometimes known as concentric folds since they can be formed by a concentric series of circular arcs. However, this is rather unusual, so that they are better termed parallel folds. It is a characteristic feature of parallel folds that the folded surfaces do not have the same shape throughout the fold. Thus, these surfaces increase in curvature towards the fold core in such a way that the folding is disharmonic.

Similar folds differ from parallel folds in that they are defined as folds in which the thickness *T* of the folded layers remains constant, parallel to the direction of the axial planes, as shown in Figure 3.16B. This means that the thickenss of each layer, as measured normal to the folded surfaces, decreases from a maximum at the fold hinge to a minimum on the fold limbs. It is a characteristic feature of similar folds the surfaces affected by the folding are identical in shape. Accordingly, the curvature shown by each surface varies in exactly the same way. In theory, such folds maintain the same profile for an infinite distance, parallel to their axial planes.

Variation in Thickness of a Folded Layer.

This discussion has introduced various aspects of fold morphology, which can now be defined with greater precision. If the profile of a cylindroidal fold is considered, parallel lines *AB* and *CD* can be drawn in pairs so that they are tangential to the surfaces forming the boundaries of a folded layer, as shown in Figure 3.17. The attitude of these lines can be described the angle *α* which they make with a line *EG* drawn at right angles

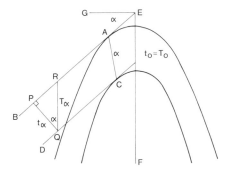

FIG. 3.17. Diagram showing how the thickness t_α and T_α of a folded layer can be measured (after Ramsay).

to the axial trace *EF* of the fold. Note that the angle α increases from zero at the fold hinge to a maximum on the fold limbs. The orthogonal thickness t_α is then defined as the perpendicular distance *PQ* between the parallel lines drawn as tangents to the folded surfaces at an angle α. Similarly, the thickness T_α of the layer is given by the distance *QR* between these lines, as measured parallel to the axial plane of the fold. Figure 3.17 shows that t_α and T_α are measured along the sides of a right-angled triangle *PQR* so that they are related according to the equation $t_\alpha = T_\alpha \cos \alpha$.

In general, the thickness of a folded layer changes around the fold hinge, so that the parameters t_α and T_α vary with the angle α. This variation can be described by defining two ratios t_α/t_0 and $T_\alpha T_0$, such that t_0 and T_0 give the thickness of the layer at the fold hinge where $\alpha = 0°$. The orthogonal thickness t_0 at the fold hinge is measured parallel to the axial plane, so that $t_0 = T_0$. How the folded layer varies in thickness can then be shown graphically by plotting the ratios t_α/t_0 and T_α/t_0 against the angle α, in the manner of Figure 3.18.

Each graph is divided into three fields according to how the ratios t_α/t_0 and $T_\alpha T_0$ vary with the angle α. Thus, if the folds are similar in style, the thickness T_α parallel to the axial plane remains the same, so that $T_\alpha/T_0 = 1$, while the orthogonal thickness t α normal to the folded surfaces varies with the angle α, so that $t_\alpha/t_0 = \cos \alpha$. The lines corresponding to these equations are plotted on the appropriate graph, in order to divide Class 1 folds from Class 3 folds. Thus, similar folds are known on this classification as Class 2 folds. Likewise, if the folding is parallel in style, the orthogonal thickness t_α normal to the folded surfaces remains the same, so that $t_\alpha/t_0 = 1$, while the thickness T_α parallel to the axial plane varies with the angle α, so that $T_\alpha/T_0 = \sec \alpha$. The lines corresponding to these equations are plotted on the appropriate graph, in order to divide Class 1A folds from Class 1C folds. Thus, parallel folds are known on this classification as

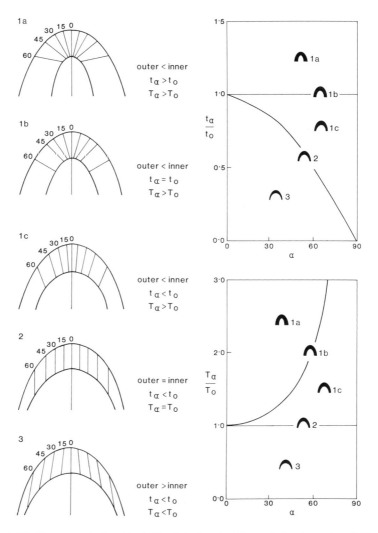

FIG. 3.18. *Upper graph:* the ratio t_α/t_o plotted against the angle for the various folds as shown to the left. *Lower graph:* the ratio T_α/T_o plotted against the angle for the various folds as shown to the left.

Class 1B folds. This means that the ratios t_α/t_0 and T_α/T_0 vary with the type of folding as follows:

Class 1A folds	$t_\alpha/t_0 > 1$	$T_\alpha/T_0 > \sec \alpha$
Class 1B Parallel folds	$t_\alpha/t_0 = 1$	$T_\alpha/T_0 = \sec \alpha$
Class 1C folds	$1 > t_\alpha/t_0 > \cos \alpha$	$\sec \alpha > T_\alpha/T_0 > 1$
Class 2 similar folds	$t_\alpha/t_0 = \cos \alpha$	$T_\alpha/T_0 = 1$
Class 3 folds	$t_\alpha/t_0 > \cos \alpha$	$T_\alpha/T_0 < 1$

This classification can be applied to any natural fold, simply by measuring the ratios t_α/t_0 and T_α/T_0 for various values of the angle α and plotting the results graphically.

Use of Dip Isogons.

The parallel lines used to define these ratios are tangential to the surfaces forming the boundaries of the folded layer wherever these surfaces have the same slope. Accordingly, a straight line AC can be drawn across the layer to join these points of equal dip, as shown in Figure 3.17. Such a line is known as a dip isogon. A series of dip isogons can be constructed for an individual fold so that each line corresponds to a different value for the angle α. For example, dip isogons can be drawn at 15° intervals, so that the angle α would have values of 0°, 15°, 30°, 45° and so on. This can be done very simply. A straight line is drawn on the fold profile, parallel to the axial plane of a particular fold. A protractor and a set-square are then used to draw a pair of parallel lines at an angle 90° - α to the axial plane so that these lines are tangential to the folded surfaces. Finally, a straight line is drawn across the layer to join these points of equal dip, where the parallel lines touch the folded surfaces tangentially. The values for t_α and T_α corresponding to a particular value for the angle α can also be measured, once this has been done. The procedure is then repeated for the various values of the angle α, as required. These dip isogons will then form a pattern which is diagnostic of the fold class, as shown in Figure 3.18.

In general, a folded layer varies in thickness according to how the surfaces forming its boundaries vary in curvature with respect to one another. Assume, for example, that the surface forming the outer arc of a folded layer always has the same shape. If the inner arc is identical in form, so that the curvature is the same at corresponding points on each surface, the points of equal dip will lie at the same distance from the axial plane. This means that the dip isogons would be parallel to one another. Since the thickness T_α of the layer parallel to the axial plane remains constant, this pattern is typical of Class 2 similar folds. However, this is obviously a special case since the inner arc might show a greater or lesser degree of curvature than the outer arc.

If the inner arc had a greater degree of curvature, it would change more rapidly in slope than the outer arc. Since the points of equal dip would be closer to the axial plane on the inner arc, in comparison with the outer arc, the dip isogons would converge on the axial plane as they are traced towards the fold core. Since the thickness T_α of the layer parallel to the axial plane would then be greater on the fold limbs than its value T_0 at the fold hinge, this pattern is typical of Class 1 folds. A special case can be recognised wherever the dip isogons converge on the axial plane so that they are always perpendicular to the folded surfaces. Since the orthogonal thickness t_α of the layer remains the same under these circumstances, this pattern is typical of Class 1B parallel folds.

This means that Class 1A folds can be distinguished from Class 1C folds according to how the dip isogons converge on the axial plane. Thus, if they converge rapidly, so that the dip isogons are inclined towards the axial plane at an angle which exceeds the outward dip of the folded surfaces, the fold belongs to Class 1A. However, if the dip isogons converge slowly, so that they are inclined towards the axial plane at an angle which is less than the outward dip of the folded surfaces, the fold belongs to Class 1C.

Alternatively, if the inner arc has a lesser degree of curvature than the outer arc, it would change more slowly in slope. Since the points of equal dip would be closer to the axial plane on the outer arc, in comparison with the inner arc, the dip isogons would diverge from the axial plane as they are traced towards the fold core. Since the thickness T_α of the layer parallel to the axial plane would then be less on the fold limbs than its value T_0 at the fold hinge, this pattern is typical of Class 3 folds.

Composite Similar Folds.

This classification only considers how a single layer varies in thickness as a result of the folding. However, if dip isogons are constructed through a series of folded layers, an important feature of fold morphology can often be discerned. If the dip isogons are parallel to one another throughout the fold, so each layer has the form of a Class 2 fold, all the folded surfaces are identical in form. It has already been mentioned that such a fold could maintain the same profile forever along its axial plane, if this were possible. However, Class 1 folds and Class 3 folds must change in shape along their axial planes. Thus, these folds become narrower and tighter as the dip isogons converge on one another in a particular direction, whereas they become broader and more open as the dip isogons diverge from one another in the opposite direction. This means that folds of this type tend to be disharmonic, dying out along their axial planes, unles they change into Class 2 folds in this direction. However, Class 1 folds and Class 3 folds are often found associated with one another in such a way that the form of the fold profile is preserved along the axial plane. Such folds are known as composite similar folds (see Fig. 3.19). They are commonly developed wherever similar folding affects layers which differ in mechanical competence. Thus, the competent layers more resistant to deformation tend to form Class 1B folds, whereas the incompetent layers less resistant to deformation thicken into the fold hinges to form Class 3 folds in such a way that the overall form of a Class 2 fold is preserved.

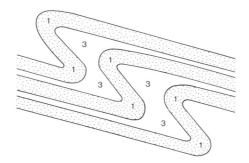

FIG. 3.19. Diagram of a composite similar folding showing how the competent layers (stippled) are affected by class 1 folds while the incompetent layers thicken into the fold hinges to form Class 3 folds.

Mechanisms of Folding

It is a reasonable assumption that the folding mostly occurs as a result of lateral compression, since it has the effect of reducing the overall length of the folded layers. This is particularly clear in the case of parallel folds, which are likely to preserve the original length of the folded layers, as measured along the arcs formed by these layers in their folded state. If this were not the case, it would be difficult to understand why the layers should have the same orthogonal thickness in such a fold. However, lateral compression is likely to be associated with the development of non-parallel folds, in which the orthogonal thickness of the individual layers is altered by the folding. Thus, such folds are commonly accompanied by the development of compressional features such as slaty cleavage, which forms parallel to their axial plane.

It is the layered nature of sedimentary rocks which allows folds to form as a result of lateral compression. This is expressed by the lithological differences which are developed between the various beds of a sedimentary sequence. It might be expected that some layers would tend to resist any lateral compression, while other layers would be more susceptible to the deformation. The layers offering the most resistance to the deformation are known as competent layers, while the layers offering the least resistance are termed incompetent.

It is the competent layers which tend to control the folding of the sedimentary sequence as a whole. The incompetent layers simply deform so that the folds developed in the competent layers can be accommodated. The competent layers of a sedimentary sequence are usually formed by sandstones and limestones, while the incompetent layers are formed by shales and evaporites. There is usually a complete gradation in competency between these rock types, so that some layers are relatively competent whereas other layers are relatively incompetent. This serves to emphasise that competency is only a relative measure of the resistence offered by a particular layer to deformation. Such a layer may be regarded as competent if it is surrounded by less competent material, whereas it would be regarded as incompetent if it was surrounded by more competent material. It is the difference in relative competency between the various layers of a sedimentary sequence which determines its response to lateral compression.

Buckling of a Competent Layer.

Buckles tend to develop in a competent layer embedded in less competent material as a result of lateral compression as shown in Figure 3.20A. This may be treated as an elastic problem, if the competent layer can be considered to act as an elastic plate surrounded by viscous material. Such a system would tend to buckle into a series of sinusoidal waves, once the com-

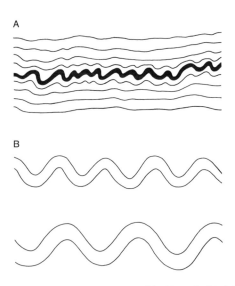

FIG. 3.20. A: Buckling of a competent layer (black) embedded in less competent material. Note how the amplitude of the folds decreases away from the competent layer. B: Diagram showing how the wavelength of a buckle fold depends on the thickness of the layer.

pressive force acting along the plate exceeds a certain value. Alternatively, it can be assumed that the layers behave in a viscous manner over the length of time that is needed for the folds to develop. This means in effect that they will deform permanently by flowing under the slightest compression.

Under these circumstances, any irregularities in the original configuration of the competent layer will tend to be amplified as the result of lateral compression, even if it does not buckle. These irregularities can be considered to form a whole series of sinusoidal waves of low amplitude, differing in wavelength from one another. They grow at a rate which varies according to their wavelength. Thus, it is only certain wavelengths which are selectively amplified to form folds in the competent layer. These folds have a wavelength which is directly proportional to the thickness of the competent layer, assuming that the difference in competency between the folded layer and the surrounding material remains a constant, as shown in Figure 3.20B. The wavelength of the dominant folds also increases as such a difference in competency becomes greater. These folds grow in an exponential manner, since they undergo a degree of amplification which increases very rapidly after a certain amount of lateral compression has taken place.

This stage of very rapid folding may be regarded as a form of buckling. It only occurs if there is quite a marked difference in competency between the layer and the surrounding material. Otherwise, the layer would become shortened as the result of lateral compression, without the development of any folds. This process is loosely known as flattening. It is accompanied by extension at right angles to the layer, which would increase in thickness as a result. This allows buckling to be distinguished as a particular type of folding which preserves the original arc-length of the folded layers.

This discussion has only considered a single layer embedded in less competent material. However, if more than one layer of competent material is involved, it is found that these layers tend to behave independently of one another if they are more than a single wavelength apart. If this is not the case, the whole sequence tends to act in a more complex manner as a single unit, which is called a multilayer. This allows folding to occur on a larger scale.

Neutral-surface Folding.

The buckling of a competent layer surrounded by less competent material is likely to result in a process known as neutral-surface folding. The deformation affecting such a layer is concentrated at the fold hinges in such a way that the convex arc on the outside is extended while the concave arc on the inside is compressed, as shown in Figure 3.21A. Thus, the outer surface is lengthened, while the inner surface is shortened, as the layer is traced around the fold hinge. This means that there is a surface within the layer which is neither lengthened nor shortened. It is called the neutral surface.

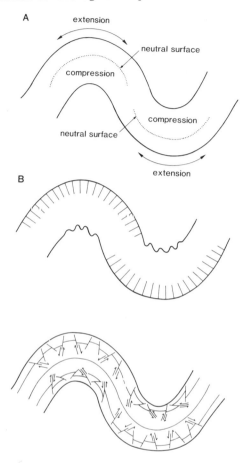

F<small>IG</small>. 3.21. A: Diagram of a neutral-surface fold. B: Compressional and extensional features associated with neutral surface folding.

The folding causes the layer to become thinner on the outer side, but thicker on the inner side, of the neutral surface. If these processes counterbalance one another, so that the layer retains its original thickness around the fold hinge, the fold would belong to Class 1B. Otherwise. a Class 1A fold would be formed by excessive thinning, whereas a Class 1C fold would be formed by excessive thickening, depending on the relative position of the neutral surface within the folded layer.

The structures associated with the developent of neutral-surface folds tend to occur at the fold hinges, since the layering on the fold limbs is not affected by any deformation, as shown in Figure 3.21A. Extension around the outer arc of a fold hinge may be reflected in the development of joints and faults, as described in Chapter 5. Extension joints would form at right angles to the folded layer, and might be filled by vein minerals, while normal faults would form at a high angle to the folded layer. Compression around the inner arc of a fold hinge may result in the development of faults and folds. Thrust faults would form at a low angle to the folded layer, while the inner arc of the folded layer may become folded on a smaller scale.

Flexural-slip Folding.

The buckling of a competent multilayer is more likely to occur by a process known as flexural-slip folding, wherever the individual layers forming such a multilayer can slip over one another. As these layers become folded, the outer layers slip over the inner layers so that they are displaced towards the hinge zones of the folds as shown in Figure 3.22A. The displacement is

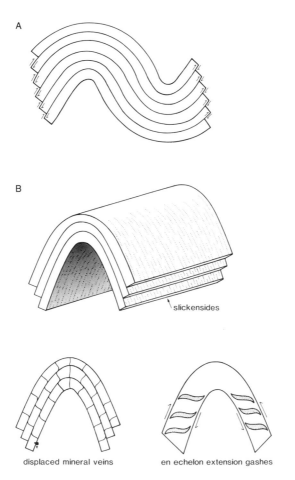

slickensides

displaced mineral veins   en echelon extension gashes

FIG. 3.22. A: Diagram of a flexural-slip fold. B: Structural features associated with flexural-slip folding.

greatest on the fold limbs, while it becomes zero at the fold hinges. The individual layers affected by flexural slip in this way must undergo neutral-surface folding as they are traced around the fold hinges. Thus, the buckling of a multilayer occurs as a combination of flexural slip between the individual layers, and neutral-surface folding which allows each layer to become folded as a result. The amount of neutral-surface folding required to accommodate flexural slip in this way depends on the relative thickness of the individual layers. It is commonly found that the multilayer is formed by a large number of relatively thin layers, so that only a slight amount of neutral-surface folding would affect each layer. Under these circumstances, the flexural-slip folding of the multilayer would lead in effect to the development of Class 1B folds, as can be seen by bending a pack of thick cards.

It is possible that perfect cohesion might exist between the layers, so that they could not slip over one another. It has been argued that the deformation would then be distributed uniformly throughout each layer. This process is known as flexural-flow folding. It can be illustrated by bending a telephone directory with very thin pages. No neutral-surface folding is required under these circumstances. Instead, the layers as a whole would be affected by a shearing deformation which increases from zero at the fold hinges to a maximum on the fold limbs.

It is unlikely that flexural flow would affect the competent layers in a multilayered sequence to any extent. Thus, flexural-flow folding requires the deformation to extend throughout the fold limbs, whereas neutral-surface folding concentrates the deformation at the fold hinges. Less energy would probably be expended in the former case, so that neutral surface folding is favoured. However, the incompetent layers are quite likely to be affected by flexural flow, as the competent layers undergo neutral-surface folding.

The structures associated with flexural-slip folding are found on the fold limbs. They are chiefly represented by the development of bedding-plane thrusts between the competent layers. These thrusts may displace early formed mineral veins as they are traced across the folded layers, as shown in Figure 3.22B. Any slickensides developed on these bedding-plane thrusts tend to lie at right angles to the plunge of the fold hinges. Extension joints may also be produced by the shearing force which affects each layer on the fold limbs, as the competent layers slip over one another during flexural-slip folding. These joints tend to form at 45° to the boundaries of the folded layer, at such an attitude that they dip towards the axial planes of the antiforms. If the layer is affected by a certain amount of flexural flow, they may open to form *en echelon* zones of sigmoidal extension-gashes, filled with vein minerals such as quartz, calcite and dolomite.

Chevron folding.

Although parallel folds formed by flexural slip often have rather rounded hinges, this is not always the case. The same mechanism can produce parallel folds with straight limbs and rather angular hinges. Such folds are known as chevron folds if they are developed on a large scale (see Fig. 3.23). They typically occur in well-bedded sequences of sandstones, where each bed has approximately the same thickness as its neighbours. These competent layers may be separated from one another by a certain amount of shale, which acts in an incompetent manner. It is the competent layers of sandstone which are affected by flexural-slip folding to form a series of chevron folds.

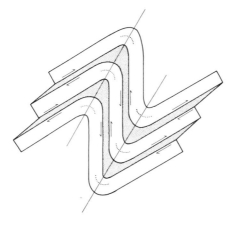

FIG. 3.23. Diagram of a chevron fold, showing how the incompetent layers are affected by flexural flow-folding while the intervening layers of competent rock are affected by neutral-surface folding, as indicated by the dotted lines.

It is a characteristic feature of chevron folds that they maintain the same shape along their axial planes. This means that they have the overall form of similar folds. However, this occurs as a result of flexural-slip folding. The competent layers are affected internally by neutral-surface folding. If there is little or no shaly material between the competent layers, flexural slip tends to occur between these layers. This often leads to cavities opening at the fold hinges, which are filled with minerals such as quartz and calcite to form saddle reefs (see Chapter 5). However, if there is shaly material between the competent layers, it would be affected by flexural flow. This means that the competent layers are folded so that they more-or-less retain their original thickness in the form of a Class 1B parallel fold, while the incompetent material thickens into the fold hinges in the form of a Class 3 fold. It is possible that chevron folding occurs as a result of buckling, although this is not certain by any means. If so, chevron folds would be formed with their axial planes at right angles to the direction of lateral compression. This may be compared with the case of kink folding, to be discussed at the end of this section.

Development of Composite Similar Folds.

Neutral-surface and flexural-slip folding tends to occur under non-metamorphic conditions which do not allow any recrystallisation or mineral growth to take place. However, there is a limited amount of shortening that can be accommodated by folding of this sort, which leads in general to the development of Class 1B parallel folds. If these folds are formed by a concentric series of circular arcs, it can easily be shown that the folding will result in a maximum shortening of 36 per cent, which occurs when the limbs have rotated into parallelism with one another. Any further deformation cannot be accommodated by the folding. However, the folds produced by neutral surface and flexural-slip folding may be affected by flattening, once this point of maximum shortening has been reached.

Such flattening would obviously modify the parallel form of a Class 1B fold. The effect can be illustrated by considering the dip isogons drawn across a folded layer, which is assumed to be bounded by circular arcs forming a parallel fold, as shown in Figure 3.24. The flattening would cause

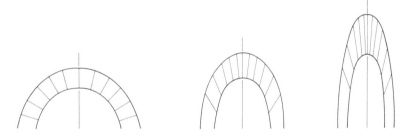

FIG. 3.24. Diagram showing how the shape of a Class 1B parallel fold (*left*) can be modified by flattening normal to its axial plane to form Class 1C folds (*right*). The form of a Class 2 similar fold would be produced after an infinite amount of flattening.

these arcs to become elliptical in shape. The dip isogons would then form a convergent pattern typical of Class 1C folds. Moreover, as the amount of flattening increases, these dip isogons would gradually approach a pattern typical of Class 2 folds, in which they would be parallel to one another. However, this situation would only be reached after an infinite amount of deformation has taken place. This means that a flattening deformation modifies parallel folds so that they become more like similar folds.

This process is only likely to occur during regional metamorphism, which would allow the deformation to be accompanied by recrystallisation and mineral growth. The temperature and confining pressure would then be sufficient for the rocks to deform in a ductile manner. However, it may be argued that it is differences in competency which cause a layered sequence to buckle under lateral compression. This means that a uniform flattening is unlikely to affect this sequence after these buckles had developed into parallel folds. Instead, it is more likely that flattening would occur as the layers started to buckle. In fact, there is usually a certain amount of layer shortening developed by lateral compression, which occurs before buckling becomes important. This would continue as a flattening deformation, which modifies the folds formed by buckling as soon as they start to develop in a competent layer. In other words, bucking and flattening occur as simultaneous processes, leading to the development of folds under metamorphic conditions.

If flattening does modify the buckling process in this way, composite similar folds would be formed under most circumstances. The competent layers buckle to form Class 1C folds, which are affected by a certain amount of flattening. The incompetent layers take on the form of Class 3 folds in order to accommodate the folding of the competent layers. The folding shows a marked contrast in style wherever there are marked differences in competency between these layers. The folding of a competent layer comes to resemble a Class 1B fold most closely as the difference in competency becomes greater between such a layer and its surroundings. This means that similar folds tend to be formed wherever there are only slight differences in competency between the various layers. Even so, it is still found that there are slight differences in fold style between the layers. Some layers take up the form of Class 1C folds, while other layers have the form of Class 3 folds, even although these folds do not depart greatly from the shape of Class 2 folds. Such folds tend to have regular wavelengths, which are closely related to the thickness of the layers. This suggests that they were formed by buckling, accompanied by a very considerable amount of flattening.

Development of Perfectly Similar Folds.

It is rather difficult to account for the perfect development of Class 2 similar folds. Indeed, it might be doubted if any Class 2 folds exist in nature. Most folds of a similar form are seen on investigation to depart to a certain extent from the ideal geometry of a Class 2 fold. However, if perfectly similar folds do occur, they must be developed as the result of differential flow acting across the folded layers.

Differential flow may be regarded as analogous to the laminar flow of liquids, which might occur as rocks are deformed very slowly under conditions of high-grade regional metamorphism. The flow occurs by individual particles moving along straight-line paths, which are identical for all the particles that have passed through the same point. In other words, there is no turbulent motion. The paths followed by individual particles are known as flow lines. If the flow is not uniform, a surface formed by a series of particles lying at an oblique angle to the flow lines would be folded. If all the flow lines were parallel to one another, a series of perfectly similar folds would be developed wherever more than one surface was affected by differential flow in this way, as shown in Figure 3.25. Thus, the folds would have axial planes which were parallel to the flow lines. The thickness T_a of a folded layer would remain constant if these flow lines were parallel to one another. It should be noted that the folded layers merely act as passive markers, which do not have any influence on the folding.

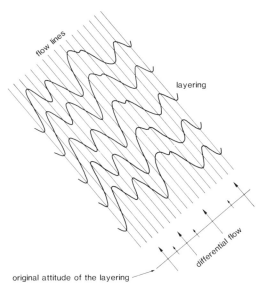

FIG. 3.25. Diagram showing the development of perfectly similar folds as the result of differential flow.

This hypothesis of differential flow-folding faces certain difficulties. If the spacing of the flow lines does not change during the deformation, there would be no compression at right angles to the axial planes of the folds so formed. However, such folding is often accompanied by the development of an axial-planar foliation, which is generally regarded as a result of compression at right angles to this direction. If this is the case, it is difficult to understand why buckling of the layers does not lead to the development of composite similar folds under these circumstances. It is necessary to assume that differential flow-folding only occurs if there are no differences in competency between the folded layers. This is unlikely to be the case under any conditions of temperature and confining pressure, unless the layering is simply defined by differences in colour between layers which are otherwise identical in lithology.

Shear folding may be distinguished as a type of differential flow-folding, which lacks any inference that the deformation occurs as a result of laminar flow. Instead, it may be assumed that slip occurs on discrete shear planes, which cut across the folded layers, separating what are known as microlithons from one another. The thickness of these microlithons depends on the spacing of the shear planes. How the folding occurs can be illustrated by shearing a pack of thick cards so that the ends take on a sigmoidal shape, as shown in Figure 3.26. The cards act as the microlithons,

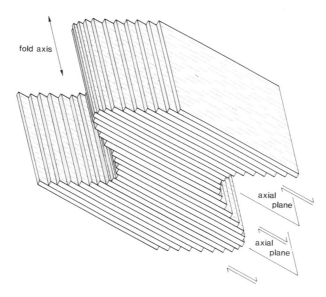

Fig. 3.26. Diagram illustrating the shear-fold hypothesis for the development of perfectly similar folds.

while the surfaces between the cards represent the shear planes. This model can also be applied to the case where the shear planes are infinitely close together, so that the layering is folded into a series of smooth curves without the development of any step-like discontinuities, as the result of differential shear.

It should be noted that the layering can be represented by planes drawn with any orientation across the pack of cards. This serves to emphasise that the layering must act in a passive manner. In particular, the fold hinges would be parallel to the direction in which the layering intersects the shear planes. It does not need to be perpendicular to the shear direction, in which the cards slip over one another (see Fig. 3.27). This can be compared to the

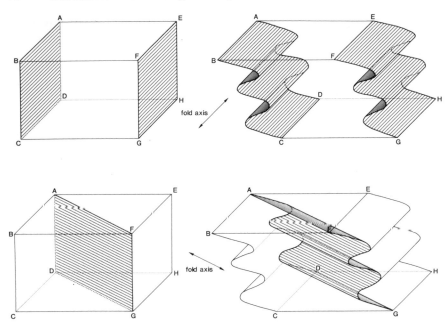

FIG. 3.27. Diagram showing how the orientation of the fold axis depends on the attitude of the layering to be folded according to the shear-fold hypothesis. Note that the fold axis is formed parallel to the direction in which the layering intersects the shear planes, as shown by the ruled lines.

case of flexural-slip folding, where the layers generally slip over one another at right angles to the fold hinges, even although this need not be the case.

Shear folding is unlikely to occur for the same reason as differential flow-folding, if it is assumed that the axial planes form at right angles to the direction of maximum compression. However, it becomes much easier to account for the development of perfectly similar folds if this assumption is not made. The direction of shear would then not need to reverse periodically on passing from one fold link to another, which is difficult to understand in mechanical terms. It is likely that such folds could be formed if differential shearing affected discrete zones within the rock-mass, lying at an moderate angle to the direction of maximum compression. Ideally, this would lead to the development of shear folds as conjugate sets, with their axial planes at a high angle to one another, even although only one set might be present under normal circumstances. The rock lying between these zones of differential shear would not be deformed to any extent. This places an important constraint on the type of deformation which can affect the intervening shear zone. It is found that perfectly similar folds would be developed, even if there were differences in competency between the layers, unless the conditions for chevron folding were satisfied.

Kink Bands and their Formation.

Angular folds with straight limbs and narrow hinges are known as kink folds, particularly if they are developed on a small scale. They commonly occur in pairs between parallel axial planes, so forming kink bands, as shown in Figure 3.28. The width of such a kink band is usually very much

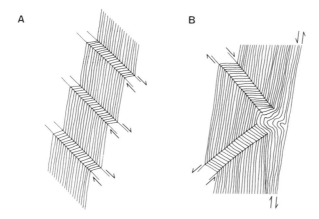

Fig. 3.28. A: Kink-folds formed by a series of kink bands cutting across finely banded layering. B: A conjugate pair of kink folds.

less than its lateral extent. This may be regarded as the characteristic feature of kink bands. Such structures are often found as conjugate sets in finely banded or well-foliated rocks. They are typically developed in low-grade metamorphic rocks, wherever deformation has produced a tectonic layering from bedding and the early foliations. These sets are usually symmetrically arranged so that they make an angle of more than 45° to the layering, which corresponds to the direction of maximum compression.

The folding occurs by flexural slip on the layering within the kink bands so that each layer preserves its original thickness. This requires that the layering inside and outside a kink band makes the same angle with the planes forming its boundaries, unless an opening is formed along the layering within the kink band. It may be assumed that each layer undergoes neutral-surface folding at the fold hinges. This can lead to the development of potential cavities along the boundaries of a kink band. If developed, such cavities would be filled by the deposition of minerals such as quartz and calcite. Since the layers affected by kink folding in this way are extremely thin, the fold hinges tend to be very angular. This means that they are likely to form planes of weakness in the rock, along which faulting can take place.

Since kink bands often occur as conjugate sets, it is commonly found that they intersect one another. This produces symmetrical folds with their axial planes at right angles to the original attitude of the layering. These structures have fold limbs which belong in effect to different sets of kink bands. It is also the case that conjugate sets of kink bands tend to be generated wherever sliding can occur along a particular horizon within the layering. This means that they tend to die out disharmonically as they are traced towards this horizon, which may be formed by a thicker and more massive layer.

Comparison with Boudinage.

While folds are usually formed wherever a layered sequence is affected by lateral compression, other structures can be produced if the layering undergoes extension in its own plane. This simply depends on the relative attitude of the layering, since compression in one direction is nearly always accompanied by extension in another direction at right angles, taking only the two-dimensional case into consideration. Such structures are chiefly represented by boudins, which are developed wherever a competent layer breaks up to form a series of sausage-shaped bodies, lying parallel to one another, as shown in Figure 3.29. In fact, the name is derived from the

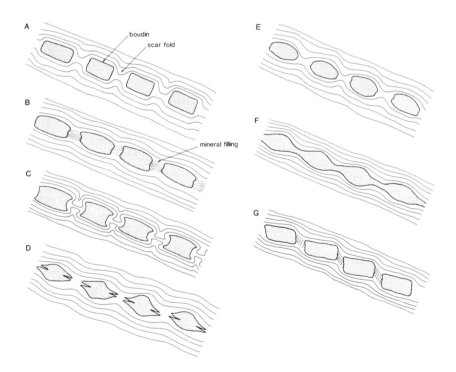

FIG. 3.29. Diagram illustrating the different types of boudinage and pinch-and-swell structures. See text for details.

French word for a blood sausage. The process whereby a competent layer becomes disrupted in this way is known as boudinage.

It is the difference in competency between the various layers which influences the shape of boudins. If a particular layer is highly competent in comparison with its surroundings, it tends to fracture on an evenly-spaced set of cross joints at an early stage in the deformation. The segments defined by these joints then move apart without much deformation affecting the layer itself. This tends to form a set of equidimensional boudins with rectangular cross-sections. (see Fig. 3.29A). The spacing of these boudins would depend on how much deformation had taken place.

By way of contrast, the competent layer may start to neck, before it ruptures to form a series of boudins, if the difference in competency is less marked. The layer is locally reduced in thickness along its length to form a series of necks, which then act as the loci for subsequent boudinage. This would produce elongate boudins with barrel-shaped cross-sections (see Figs. 3.29B and E). Such a shape may be accentuated as the incompetent material on either side of the layer flows into the spaces formed between the boudins as they move apart. As a result, the ends of the boudins can develop ears, while the original fracture becomes curved into a concave form (see Fig. 3.29C). This process may be so effective that the two ears come together to form boudins with lenticular cross-sections (see Fig. 3.29D). Finally, if there is only a slight difference in competency between a layer and its surroundings, it may only be affected by necking without any boudinage taking place. This would produce a structure known as pinch-and-swell within the layer (see Fig. 3.29F).

The flow of incompetent material into the spaces formed between the boudins forms structures which are known as scar folds, if this material is banded in any way. The hinges of these folds would be parallel to the long axes of the boudins. Alternatively, cavities may open between the boundins, which become filled with vein minerals such as quartz, calcite and dolomite, or pegmatite. Boudins may also appear to be affected by rotation, so that they take on a rhomboidal form (see Fig. 3.29G). This can occur if the layering was originally at an angle to the direction of extension. The layering would then rotate towards this direction as a result of the extension. However, it is found that the banding within the boudins tends to lag behind the rotation undergone by the layering as a whole, so that it lies closer to its original attitude. This means that it is the surrounding material rather than the boudinaged layer which is affected by most rotation under these circumstances.

CHAPTER 4

Folded Rocks and the Outcrop Pattern

Interpretation of the Outcrop Pattern

ON examining a geological map, the folding of sedimentary rocks can be recognised from its effects on the outcrop pattern of the stratigraphic formations. If the sedimentary rocks are flat-lying, it is mostly the effect of the topography on the geological boundaries which controls the nature of the outcrop pattern. However, if the sedimentary rocks have been folded, it is more the geological structure which controls the outcrop of the stratigraphic formations, unless the topographic relief is very marked. Under these circumstances, the outcrop pattern of the sedimentary formations shows how the bedding varies in strike as it is traced across the area under consideration. The dip of the bedding can be found by studying the effect of the topography on the geological boundaries, wherever this can be determined. Alternatively, the structural observations plotted on the geological map to show the dip of the bedding can be used. The form of the geological structures can then be established by studying how the stratigraphic formations vary in dip and strike as they are traced across the area under consideration.

Structure Contours on a Cylindrically Folded Structure.

An understanding of how the form of an individual fold can be determined from the structural observations plotted on a geological map is best approached through the use of structure contours. Consider first the geometry of a cylindrically folded surface, affected by a series of plunging folds. By definition, such a surface can be generated by a straight line moving parallel to itself along a curved path. Although it is generally assumed that the straight line moves in a plane at right angles to itself, this need not be the case. It could equally well follow a path corresponding to a cross-section through the folded surface in a horizontal plane, as shown in Figure 4.1. Such a horizontal cross-section is equivalent to a structure contour drawn at the same height on the folded surface. This means that such a contour could be traced out by a point inscribed at a particular height on the

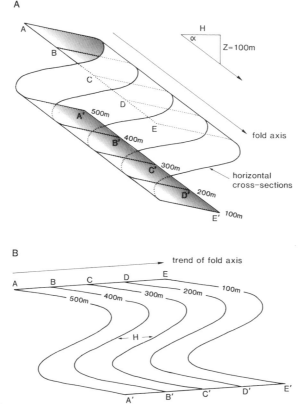

FIG. 4.1. *Above:* diagram showing a series of structure contours, *AA'*, *BB'* and so on, inscribed at a contour interval of 100 m on a cylindrically folded surface. Each contour corresponds to a horizontal cross-section through the folded surface. *Below:* plan showing the identical form and constant spacing of the structure contours drawn on this surface, parallel to the trend of the fold axis. Note that each contour line could be generated by a point such as A inscribed on a line parallel to the fold axis as it follows a path corresponding to a horizontal cross-section through the folded surface.

straight line generating the folded surface, provided that it follows a path corresponding to a horizontal cross-section through the folded surface.

Now suppose a series of points inscribed on the straight line generating the folded surface occur at such a height that they are each separated from one another by a vertical distance corresponding to the contour interval Z. The horizontal distance *H* between these points is then given by:

$$H = Z \operatorname{Cot} \alpha$$

where the fold axis plunges at an angle α. These points would then trace out identical curves, separated from one another by the horizontal distance *H* parallel to the trend of the fold axis, as the straight line moved along a horizontal path to generate the folded surface. These curves would represent an identical series of horizontal cross-sections through the folded surface, separated from one another by a vertical distance corresponding to the contour interval. This means that the structure contours drawn on a cylindrically folded surface are identical in form to one another, while they are spaced equally far apart from one another in a direction parallel to the trend of the fold axis.

Such a pattern can be generated by repeatedly displacing a contour line in a direction parallel to the trend of the fold axis through a horizontal distance h corresponding to the plunge of the fold axis. The geological boundary corresponding to the outcrop of the folded surface in areas of subdued topography can be used to define a structure contour with a particular height above Ordnance Datum. This means that structure contours can be drawn using the outcrop of a folded surface once the fold plunge has been determined. A characteristic feature of such a contour pattern is that all the contour lines are parallel to one another wherever they are crossed by a line drawn parallel to the trend of the fold axis. These contour lines generally form a series of sinuous curves or zig-zag lines, depending on the relative angularity of the fold hinges, as determined by the nature of the geological boundary representing the folded surface.

Modifications of the Contour Pattern.

It must be admitted that natural folds are rarely cylindrical, since they cannot be expected to persist forever without dying out. This means that such folds are often cylindroidal rather than cylindrical, even if they do not fall into the class of non-cylindroidal structures. Cylindroidal folds can be distinguished from cylindrical folds in that the structure contours drawn on a particular surface affected by the folding do not preserve exactly the same form at different levels of the structure, as shown in Figure 4.2A. Thus, even if the folding is cylindroidal so that the folded surface has approximately a cylindrical form, the individual contours do not remain strictly parallel to one another in the direction of the mean fold axis. The contour pattern would still resemble that developed as a result of cylindrical folding if it is assumed that all the fold hinges plunge in approximately the same direction within the area under consideration. This means that the structure contours tend to form a series of sinuous curves or zig-zag lines, which are nearly if not quite parallel to one another in the direction of the mean fold axis, as shown in Figure 4.2A.

This type of contour pattern is modified wherever a fold hinge passes through the horizontal to plunge in the opposite direction (see Fig. 4.2B). Culminations and depressions are then developed in the folded surface, giving rise to folds with the form of brachy-antiforms and brachy-

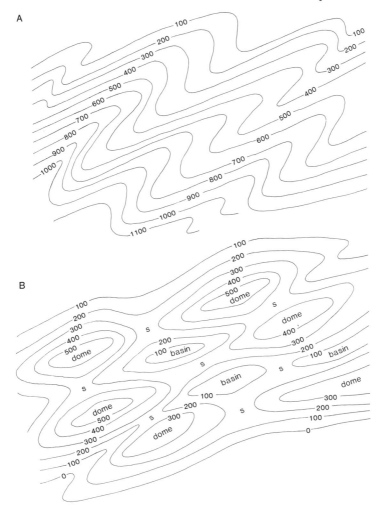

FIG. 4.2. A: Map of structure contours drawn on a cylindroidally folded surface, affected by a series of plunging folds. B: Map of structure contours drawn on a non-cylindroial surface affected by a series of culminations and depressions. S denotes saddle-shaped surfaces juxtaposed between domes and basins.

synforms. The structure contours drawn on such a surface are formed by more-or-less elliptical lines, which are each circumscribed about the core of a brachy-antiform or a brachy-synform, as the case may be. Such a contour pattern becomes more equidimensional as these elongate folds are replaced by domes and basins in the folded surface. The juxaposition of the closed elements in the contour pattern defines the saddle-shaped areas in the folded surface which separate these periclinal structures from one another.

Finally, it can be noted that horizontal folds do not result in contour patterns of the form already described for plunging folds, even although they may be cylindrical in character. If the fold axis is horizontal, this direction must be shared in common with both sets of fold limbs. This means in effect that the fold limbs must strike parallel to one another. The structure contours drawn on such a folded surface are therefore parallel to one another. This may be regarded as an extreme form of the contour pattern developed for plunging folds, in which the structure contours converge on one another at infinity as they are traced along the fold limbs towards the fold hinge. The spacing of the contour lines in a direction parallel to the fold axis is likewise at infinity, corresponding to a horizontal angle of plunge for the fold axis.

Apparent Dip and Fold Plunge.

The definition of a cylindrical fold given in the last chapter has an important corollary concerning the geometry of a cylindrically folded surface. It implies that a plane can always be drawn so that it comes into tangential contact with the folded surface along a straight line which is parallel to the fold axis, as shown in Figure 4.3A. Such a plane can be considered to define an infinitesimally small element of the folded surface in the immediate vicinity of this line.

Obviously, an infinite number of planes can be drawn so that they are all tangential to the folded surface. They would each define an infinitesimally small element of the folded surface as it changed in attitude around the fold hinges. Since all these planar elements are tangential to the folded surface, they each contain a line which is parallel to the fold axis. This line can be considered to define an apparent dip. All the planar elements of cylindrically folded surface therefore share an apparent dip in common, even although they do not have the same dip and strike as one another. This common direction is parallel to the plunge of the fold axis. These planar elements of a cylindrically folded surface also intersect one another in this direction of apparent dip, parallel to the plunge of the fold axis.

Such a planar element of a cylindrically folded surface can be identified by drawing a pair of straight lines *AC* and *BD* parallel to one another in such a way that they are each tangential to a structure contour drawn on the folded surface, as shown in Figure 4.3B. These straight lines then form another set of structure contours, drawn on a planar element of the folded strike with a particular strike. The attitude of the folded surface can therefore be found at a particular point from the trend and relative spacing of these contour lines, wherever they are tangential to the structure contours representing the folded surface itself.

This construction can be extended to find the plunge of the fold axis by drawing in a straight line *AB* through all the points where the two sets of

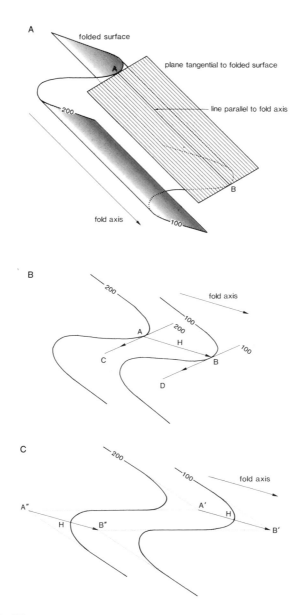

FIG. 4.3. A: Diagram showing how a planar element AB of a folded surface can be identified where it comes into tangential contact with a plane along a line parallel to the fold axis. B: Construction to determine the attitude of a planar element of a folded surface, knowing its strike as given by the parallel lines AC and BD. The plunge of the fold axis is then given by the line AB. C: Alternative construction to determine the plunge of the fold axis from a pair of structure contours, drawn on a cylindrically folded surface.

structure contours are tangential to one another. The trend of this line gives the direction of apparent dip within the folded surface, parallel to the trend of the fold axis. The intersections made by this line with the structure contours drawn on the folded surface then allow points of known height to be identified along this line. The spacing H of these points gives the angle α of plunge for the fold axis according to the equation tan $\alpha = H/Z$, where Z is the contour interval. This is equivalent to finding the angle of apparent dip within the folded surface, parallel to the plunge of the fold axis.

Alternatively, the structure contours drawn at an equivalent height on the opposite limbs of a plunging fold may be projected until they meet one another in a series of points near the fold hinge, as shown in Figure 4.3C. This effectively reconstructs an angular fold with the same plunge as the fold under consideration. The points where structure contours of the same height intersect one another then lie along a straight line which is parallel to the hinge of this fold. The bearing of this line gives the trend of the fold axis, while the spacing of points of known height along this line can be used to determine the plunge of the fold axis, according to the equation already given.

This construction can be used even if structure contours have not been drawn on a folded surface, provided that the dip of the fold limbs is known. Thus, if the topography is horizontal or nearly so, the outcrop of a folded surface defines the structure contour drawn at the same height on the folded surface, at least approximately. The other contours can then be constructed as a series of straight lines drawn parallel to the outcrop of the fold limbs, with a spacing corresponding to their respective angles of dip. These lines are then projected so that they meet one another along a line parallel to the fold axis, so that its plunge can be determined.

A similar construction can be used to determine the plunge of the fold axis from the dip and strike of a folded surface at two points. For each set of observations, a pair of straight lines is drawn parallel to the strike with a spacing that corresponds to the dip of the folded surface. A straight line is then drawn through the points in which these straight lines intersect one another to form a line parallel to the fold axis. The plunge of the fold axis can then be determined, as described in the last paragraph.

This construction illustrates an important feature of cylindrical folds, which can be used to determine the plunge of the fold axis wherever a folded surface passes through the vertical to become overturned. It has already been described how all the planar elements defining a folded surface share an apparent dip in common with one another, parallel to the plunge of the fold axis. However, all the apparent dips developed by a vertical plane are directed along its strike. This means that the trend of the fold axis is given by the strike of the folded surface wherever it is vertical. Now consider the folded surface where it strikes at right angles to this direction. The apparent dip parallel to the fold axis would then be directed at right angles to the strike of this surface, down its dip. The plunge of the fold axis is therefore given by the dip of the folded surface wherever it strikes at right angles to the vertical, as shown in Figure 4.4.

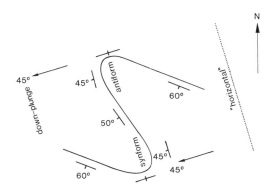

FIG. 4.4 Map showing a fold plunging at 45° towards 255°. Note that the strike where the folded surface is vertical is parallel to the trend of the fold axis, while the angle of dip of the folded surface where it strikes at right angles to this direction gives the angle of plunge. Viewing this diagram at 45° in a direction parallel to the fold axis provides a down-plunge view of the outcrop, corresponding to the fold profile.

Down-plunge View of the Outcrop Pattern.

Once the plunge of a cylindrical fold has been determined, its form can be visualised by viewing the outcrop pattern at an oblique angle, corresponding to the plunge of the fold axis. This provides a down-plunge view of the fold profile, as seen in a plane at right angles to the fold axis. This method can only be used in areas of low relief where the topography has little or no effect on the outcrop pattern. It is similar in concept to the down-dip method of viewing the outcrop patern of tilted rocks, which all dip in the same direction. However, the present method is based on the fact that cylindrically folded surfaces all share an apparent dip in common with one another, parallel to the plunge of the fold axis. By viewing the outcrop pattern obliquely in this direction, the distortion introduced by a horizontal cross-section through a plunging fold can be removed, to provide a foreshortened view of the fold profile.

The geological map should first be turned until the fold axis, which must be known at least approximately, plunges away from the observer, as shown in Figure 4.4. The outcrop pattern is then viewed at an oblique angle, corresponding to the plunge of the fold axis. This can be done by holding the geological map so that the line of'sight is horizontal, even although this is not strictly necessary. The fold profile can then be seen at right angles to the line of sight. It is important to realise that the horizontal in the plane of the fold profile is given by any line inscribed on the geological map at right angles to the plunge of the fold axis. This line can therefore be used to determine whether a particular fold closes upwards or downwards in the plane of the fold profile, so allowing antiforms to be distinguished from synforms. It should be emphasised that the form of the fold profile depends on the direction in which the outcrop pattern is viewed. For example, a fold plunging in one direction would appear to be an antiform if it is viewed in this direction, whereas it would appear to be a synform if it is viewed in the opposite direction.

How such a set of plunging folds outcrops in an area of low topographic relief can be seen from Figure 4.5. The fold limbs can be identified wherever the geological boundaries maintain the same trend for some distance along

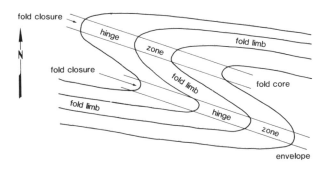

Fig. 4.5. Map showing the outcrop pattern typically developed by a series of plunging folds in an area of low relief.

the strike, while the fold hinges can be recognised wherever the geological boundaries show a marked change in strike as they are traced from one fold limb to another. Such a change in strike defines the closure of a fold, as it is seen in a horizontal cross-section. The core of the fold always outcrops on the concave side of a fold closure, while the rocks forming its envelope always outcrop on the convex side.

Axial Traces and Fold Closures.

It should be realised that the position of the fold hinges may be rather difficult to locate, since the geological map provides a distorted view of the fold profile. In particular, the point of greatest curvature in the profile of a folded surface does not always correspond to the point of greatest curvature along the outcrop of this surface. Such distortion occurs wherever the fold pitches at an oblique angle within an inclined axial plane. The exact position of the fold hinges cannot then be found by simple inspection of the outcrop pattern. However, they can be located by viewing the outcrop pattern in a down-plunge direction, since this corrects the distortion introduced by a horizontal cross-section through a plunging fold (see Fig. 4.6). The true position of successive fold hinges can therefore be located along the closure of a particular fold, so allowing the axial trace to be drawn in as a line passing through all these points.

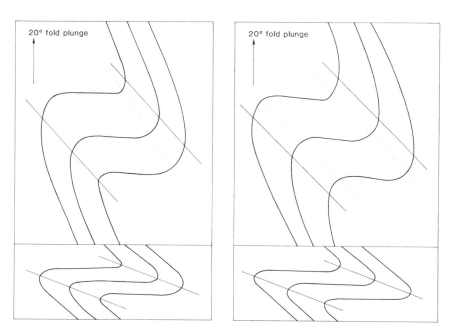

FIG. 4.6. Diagram showing how the axial traces of plunging folds (solid lines) do not correspond to their apparent position as shown in plan (dotted lines).

The effect of topography on the outcrop pattern should also be taken into account in any attempt to locate the fold closures shown on a geological map. Thus, geological boundaries show changes in trend as they cross topographic features such as ridges and valleys, even although they represent the outcrop of uniformly dipping surfaces. Such changes in trend cannot be identified as fold closures. Fold closures can therefore only be recognised where any change in trend shown by a geological boundary cannot be related to the effect of the topography on the outcrop pattern. Such changes in trend only occur where there are no topographic features present to account for their development.

It should be emphasised that the form of a fold closure on sloping ground does not reflect the form of the fold itself. Thus, a fold closing upwards on a hillside would only represent an antiform if it has a plunge close to the horizontal, if it plunges into the hill itself, or if it plunges away from the hill at a shallower angle than the hillside. Such a fold plunging more steeply in the same direction as the slope of the ground would be a synform. Likewise, a fold closing downwards on a hillside only represents a synform if it did not plunge more steeply in the same direction as the slope of the ground. Whether a particular fold closure represents an antiform or a synform therefore depends on the plunge of the fold axis in relation to the slope of the ground.

Cylindrical and Cylindroidal Folds

The form of a cylindrically folded surface can be determined from its outcrop if the plunge of the fold axis can be found using the methods already described. Consider a cylindrical fold closing towards the north-west in an area where the topography has virtually no effect on the outcrop pattern. The axial trace trends approximately from north-west to south-east, while it may be assumed for convenience that the fold limbs strike north-north-west and west-north-west respectively, to give an outcrop pattern as shown in Figure 4.7. It is important to realise that this outcrop pattern is quite independent of the direction in which the fold plunges, since not even the dip of the axial plane can be determined. Only the strike of the folded surfaces and approximate trend of the axial plane can be found from the outcrop pattern. How the plunge of the fold axis affects the dip of the folded surfaces must therefore be considered, since this allows the plunge of the fold axis to be determined from the change in attitude of the folded surfaces as they are traced around the fold hinge.

The fold axis can plunge in any direction relative to the trend of the axial plane. However, it is only necessary to consider those cases where the fold plunges towards the northern and eastern quadrants on one side of the axial trace, so that the axial plane is either vertical or dips to the north-east. How the fold varies in form can then be investigated by generating a pattern of structure contours appropriate to the plunge of the fold axis. This allows a number of different categories to be recognised, which can be distinguished from one another according to the plunge of the fold axis in relation to the strike of the fold limbs and the trend of the axial plane. The form of the folded surfaces can then be deduced from the evidence presented by a geological map, wherever the dip of the bedding is shown by means of structural observations, using the down-plunge method of viewing the outcrop pattern.

Fold Limbs Dipping in Opposite Directions.

This category includes all those folds where the fold axis plunges in such a direction that the fold limbs never pass through the vertical to become overturned. This places an important constraint on the plunge of the fold axis, since it must trend closer to the axial trace than the strike of the fold limbs. It is then found that the structure contours form a pattern in which the in-

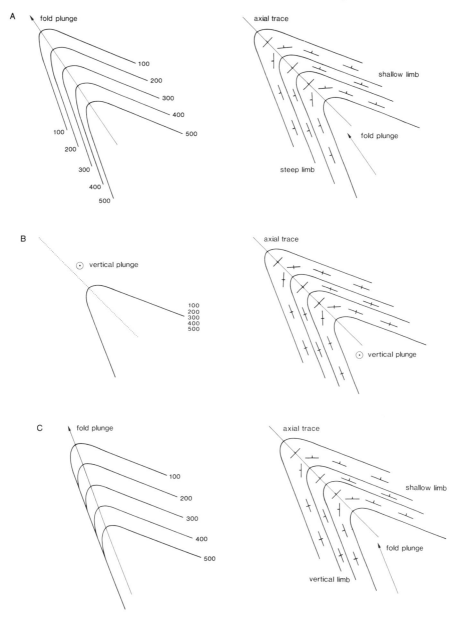

FIG. 4.7. Diagram illustrating how the dip of the folded surfaces varies as the plunge of the fold axis changes in relation to the axial trace of the fold. The diagrams on the left give the pattern of structure contours for each fold while the diagrams on the right show the corresponding dips of the folded surfaces. A: Fold with limbs dipping in opposite direction. B: Vertical fold. C: Fold with a vertical fold limb. D: Fold with an over-turned fold limb. E: Reclined fold.

FIG. 4.7 continued on next page

FIG. 4.7. Continued

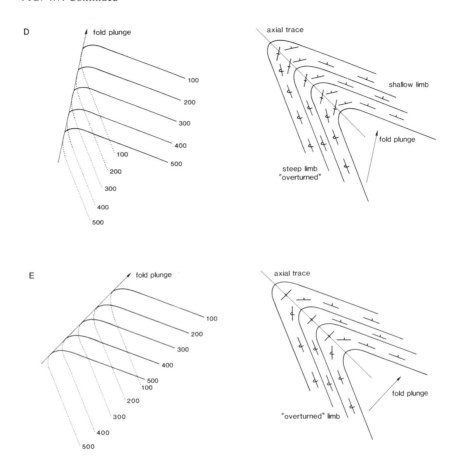

dividual contour lines do not cross one another, so that the fold limbs dip in opposite direction, as shown in Figure 4.7A. If the fold axis plunges between the north-west and the north-north-west, in the same direction as the fold closure, an antiform is present with its fold limbs dipping away from one another. The bedding at the fold hinge dips in the direction of the fold closure, away from its core. However, if the fold axis plunges between the south-east and the east-south-east, in the opposite direction to the fold closure, a synform is present with its fold limbs dipping towards one another. The bedding at the fold hinge dips towards the core of the fold, away from the fold closure. It can therefore be concluded that, if the fold limbs dip in opposite directions so that the bedding has not been over-turned, its dip at the fold hinge corresponds to the plunge of the fold hinge, at least approximately.

Vertical Folds.

If the plunge of the fold hinge increases to 90°, a vertical fold is produced as a particular type of neutral fold. Since the folded surfaces must always contain a line parallel to the fold axis, they must all be vertical. The contour pattern would then consist of a single line, representing a whole series of structure contours that are superimposed vertically above one another, as shown in Figure 4.7B. The fold limbs do not differ in dip from one another, since they are vertical, even although they strike in different directions. This tendency is exhibited by any fold with a steeply plunging hinge. It should be noted that vertical folds cannot be separated into antiforms and synforms, since they close to one side in a horizontal direction. This is shown by the fact that the bedding at the fold hinge is also vertical.

Folds with a Vertical Fold Limb.

The next category to be considered is formed by folds that plunge parallel to the strike of one or other fold limb. This fold limb must be vertical in order that its apparent dip parallel to the fold axis can also be parallel to its strike. It has already been shown that the plunge of the fold axis is given by the dip of the folded surface wherever it strikes at right angles to this fold limb. The structure contours form a pattern in which the individual contour lines coalesce with one another wherever the contoured surface becomes vertical as it is traced around the fold hinge, away from the other fold limb, as shown in Figure 4.7C. If the fold plunges towards the north-north-west, it forms an antiform with its vertical limb striking in the same direction. The other fold limb dips towards the north-north-east, away from the core of the fold. However, if the fold plunges towards the east-south-east, it forms a synform with its vertical limb striking in the same direction. The other fold limb dips towards the east-north-east, towards the core of the fold.

Folds with an Overturned Fold Limb.

This category includes all those folds where the fold axis plunges in such a direction that a folded surface passes through the vertical to become overturned. The fold axis must then trend at an oblique angle to the axial trace, beyond the strike of the overturned fold limb. Since the fold axis is parallel to a direction of apparent dip within the axial plane, while the axial trace corresponds to its strike, this means that the fold must have an inclined axial plane, dipping at an angle which varies according to the attitude of the fold axis. It is then found that the structure contours form a pattern in which the individual contour lines cross one another as the contoured surface is traced from the fold hinge towards the overturned fold limb, as shown in Figure 4.7D. This means that the fold limbs dip approximately in the same direction as one another, even although they differ in strike. If the fold axis plunges between the north-north-west and the north-east, an antiform is present with its steeper limb dipping towards the east-north-east. The other limb dips less steeply towards the north-north-east, away from the core of the fold. However, if the fold axis plunges between the east-south-east and the north-east, a synform is present with its steeper limb dipping towards the north-north-east, away from the core of the fold. The other limb dips less steeply towards the east-north-east. As previously, the trend of the fold axis is given by the strike of the folded surface where it is vertical, while its plunge from the horizontal is given by the dip of the folded surface wherever it strikes at right angles to this direction.

Whether the fold is an antiform or a synform can be determined by considering how the fold limbs converge on one another. This is best done by reversing the dip of the steeper fold limb, since it is this limb which passes through the vertical to become overturned away from the fold hinge. Traced in the opposite direction, this fold limb passes through the vertical to dip away from the core of the fold in the vicinity of the fold hinge, if the fold is an antiform, whereas it passes through the vertical to dip towards the core of the fold in the vicinity of the fold hinge, if the fold is a synform.

Reclined Folds.

The final category to be considered includes all those folds which plunge at right angles to the trend of the axial trace, so forming reclined folds. A direction parallel to the fold axis would then plunge down the dip of the axial plane. The structure contours form a pattern in which the individual contour lines are displaced at right angles to the axial trace, so that they cross one another in a symmetrical fashion, as shown in Figure 4.7E. It is a characteristic feature of such folds that the fold limbs dip at the same angle, even although they differ in strike, while the bedding at the fold hinges is vertical. This tendency is exhibited by any fold with a hinge that pitches at a high angle within its axial plane. Reclined folds resemble vertical folds in that they cannot be distinguished as antiforms or synforms, since they close to one side in a horizontal direction. They are therefore classified as neutral folds.

Non-Cylindroidal Folds

Another type of outcrop pattern is developed by periclinal structures of various kinds. These structures may have no particular trend, so that they occur as domes and basins, or they may be represented by doubly-plunging folds with the elongate form of brachy-anticlines and brachy-synclines, as shown in Figure 4.8. It is commonly found that individual formations can be traced around these structures, dipping outwards in the case of domes and brachy-anticlines and inwards in the case of basins and brachy-

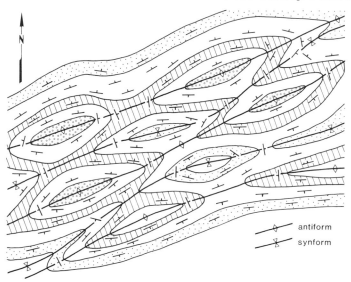

FIG. 4.8. Geological map showing the type of outcrop pattern developed by doubly plunging folds in the form of brachy-anticlines and brachy-synclines. Compare with Figure 4.2B.

synclines if the bedding has not been overturned anywhere. The sedimentary formations then form a concentric series of annular outcrops, showing circular outlines in the case of domes and basins but elliptical outlines in the case of brachy-anticlines and brachy-synclines.

It should be understood that axial traces cannot be drawn for domes and basins, unless they are not strictly equidimensional in form. However, axial traces can be drawn for brachy-aticlines and brachy-synclines, at least approximately, passing through successive fold closures. This means in effect that each outcrop ends in a fold closure where the geological boundary shows a swing in strike as it passes across the fold hinge. The bedding at the fold hinge dips away from the fold core in the case of a brachy-anticline, while it dips towards the fold core in the case of a brachy-syncline. It is the overlying formations which are encountered in the same direction as a brachy-anticline closes along its axial trace, while it is the underlying formations which are encountered in this direction in the case of a brachy-syncline.

Such an outcrop pattern is also developed wherever horizontal folds die out. Thus, the terminations of the individual folds are usually marked by a gradual increase in the plunge of the fold hinges before they eventually disappear. Antiforms are found to plunge downwards from the horizontal, while synforms show the opposite relationship. This means that horizontal folds tend to adopt the form of doubly-plunging folds if they are traced far enough in either direction, since they must eventually die out. The ends of these folds are therefore marked by a series of plunging fold closures, beyond which the structures cannot be traced. Particular attention should be paid to the dip of the bedding at these fold closures, where the geological boundaries cross the axial traces of the folds. It has already been shown that the bedding dips away from the core of an antiform in the same direction as it closes, while the bedding dips towards the core of a synform in the opposite direction to its closure, so allowing the nature of a fold closure to be determined.

Horizontal Folds

A third type of outcrop pattern is developed by cylindroidal folds with horizontal hinges. Since the fold limbs do not differ in strike from one another, the sedimentary formations generally form a series of parallel outcrops trending in the same direction as one another, at least in areas of low relief. It is essentially the repetition of the same stratigraphic sequence in reverse order which allows such an outcrop pattern to be interpreted in terms of folding. In particular, the axial traces of the individual folds can be located as lying within any formation which is flanked on either side by the same sequence of sedimentary formations in reverse order. For example, axial traces can be drawn within formations A and D wherever they are flanked on either side by formations B and C respectively, as shown in Figure 4.9A.

The form of these folds can only be determined if the dip of the bedding on the fold limbs is known. The case is relatively simple if no overturning has occurred. It would then be found that antiforms have fold limbs dipping away from one another, while synforms have fold limbs dipping towards one another, as shown in Figure 4.9B. However, if the folds are overturned, the bedding passes through the vertical to dip in the same direction on both sets of fold limbs, as shown in Figure 4.9C. Unless the folding is isoclinal, there would be a difference in the angle of dip between these fold limbs. Thus, a fold can be recognised as an antiform wherever it is the steeper limb which dips towards its axial trace, whereas such a fold can be recognised as a synform wherever the steeper limb dips away from its axial trace. This difference arises from the fact that the lower limbs of antiforms are more steeply inclined than their upper limbs, while the opposite is true of synforms.

It should be emphasised that reversals in the dip of the bedding are not always sufficient to establish the presence of folding, unless some measure of stratigraphic control is available. This is provided, for example, by the repetition of sedimentary formations as a result of the folding, which causes the same stratigraphic sequence to be encountered in reverse order as the outcrop pattern is traced across a folded terrain. The absence of such

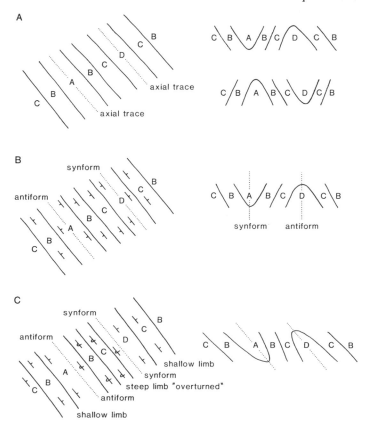

FIG. 4.9. *Left:* geological maps showing the type of outcrop pattern developed by horizontal folds in an area of low relief. *Right:* vertical cross-sections showing structural interpretations of the outcrop pattern. A: Outcrop pattern without any structural observations. B: Horizontal folds without any overturning. C: Horizontal folds with an "overturned" fold limb.

evidence allows two interpretations to be advanced wherever the bedding changes in dip so that the rocks appear to be inclined towards one another along a particular section, as shown in Figure 4.10. The first interpretation assumes that the bedding passes through the horizontal to dip in opposite directions, so that a synform is present. The second interpretation assumes that the bedding passes through the vertical to dip in opposite directions, so that the presence of a synform cannot be established.

The two cases can be distinguished if there is sufficient exposure to show whether the bedding passes through the horizontal or the vertical along the section. Alternatively, sedimentary structures may be used as a measure of stratigraphic control to show whether or not there is a change in the direction of younging along the section. The presence of a synform would re-

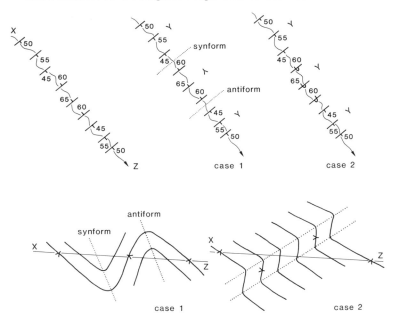

FIG. 4.10. Geological maps of a stream section showing how stratigraphic observations can be used to distinguish between two possible interpretations of the geological structure, as illustrated by vertical cross-sections.

quire that the bedding passed through the horizontal to young in opposite directions on either fold limb. Otherwise, the bedding would pass through the vertical to become inverted on the evidence of sedimentary structures, so that there would be no reversal in the direction of younging along the section. This means in effect that stratigraphic observations concerning the younging of the beds should be used, as well as structural observations concerning the dip of the bedding, to establish the presence of folds in a particular area, unless the folding is relatively open.

Anticlines and Synclines

An anticline can only be distinguished from a syncline according to whether older or younger rocks occur in the core of the fold. This may be determined in a number of different ways, which have already been discussed in Chapter 2. The Principle of Superposition can only be used if the sedimentary formations affected by the folding can be traced into areas of less complexity, where they form flat-lying sequences in order of their deposition. It may also be possible to date individual formations stratigraphically by means of fossil evidence, which would allow their stratigraphic order to be determined. Alternatively, the structural observations plotted on the geological map may show the direction of younging, which can be determined from the presence of sedimentary structures of various kinds, as well as the dip of the bedding.

All this information is likely to be incorporated into any stratigraphic column which accompanies the geological map, showing the sedimentary formations in stratigraphic order, while the structural observations concerning the younging of the beds should also be plotted on the geological map. A particular fold can then be distinguished as an anticline or a syncline, according to this evidence, as shown in Figure 4.11. The sedimen-

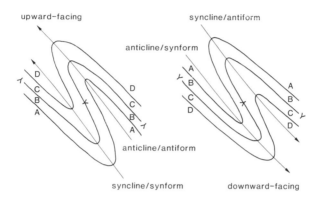

FIG. 4.11. Diagram showing that the definition of a fold as an anticline or a syncline is independent of its attitude.

tary formations would young away from the core of an anticline, whereas they would young towards the core of a syncline. The direction of structural facing may then be determined, once a particular fold has been identified as an anticline or a syncline, by comparison with its structural form as an antiform or a synform. It is often assumed that folds face upwards rather than downwards, unless there is clear evidence to the contrary. However, the limitations of such an assumption must always be borne in mind, wherever detailed stratigraphic evidence is lacking to determine the direction of structural facing, particularly in areas of structural complexity.

Construction of Fold Profiles

While the general character of a cylindroidal fold may be determined by inspection of the outcrop pattern in relation to the structural observations plotted on a geological map, only a profile drawn at right angles to the mean fold axis shows the detailed form of the fold in true perspective. The construction of such a fold profile is based on an important feature of the geometry shown in particular by cylindrical folds. It has already been emphasised that the planar elements defining a set of cylindrically folded surfaces all share a direction of apparent dip in common with one another, parallel to the fold axis. The folded surfaces can therefore be traced in the same direction, without showing any change in attitude.

Now consider a point lying on a geological boundary which represents the outcrop of a cylindrically folded surface. This point can be projected in a direction parallel to the fold axis in such a way that it still lies within this surface. This means in effect that the geological boundary can itself be projected in the same way, until it forms a line within the plane of the fold profile. This has the effect of correcting the distortion introduced by any horizontal cross-section through a cylindrical fold, so that the outcrop pattern can be converted into a fold profile, showing the true form of the fold in a plane at right angles to the fold axis. This method can obviously be applied with little error to cylindroidal folds, even although the profile would then show slight changes in form as it is traced down the plunge of the mean fold axis.

Horizontal Folds.

A special method is required for the construction of a profile across a series of horizontal folds. This method illustrates in a simple way how individual points in the outcrop pattern are projected parallel to the direction of the mean fold axis until they lie in the plane of the fold profile. The construction can only be used if the topography has a high relief. It might then be possible to contour a particular surface affected by the folding. The structure contours would be parallel to the direction of the mean fold axis. Under these circumstances, the plane of the fold profile is vertical, striking at right angles to the direction of the mean fold axis.

The method assumes that the points lying along a geological boundary can be projected in a horizontal direction parallel to the mean fold axis. Consider, for example, the point where a geological boundary crosses a topographic contour. This point is projected along the structure contour at the same height, until it lies within the plane of the fold profile. The geological boundary as shown in profile must obviously pass through this point. In other words, any point on a geological boundary can be projected parallel to the direction of the mean fold axis, so that it still lies within the folded surface represented by this boundary. This forms the basis for the construction now to be described. It has the advantage that it does not require structure contours to be drawn, even although the direction of the mean fold axis must be known.

This method requires that vertical cross-sections are drawn at a suitable interval across the geological map at right angles to the mean fold axis, as shown in Figure 4.12. Each cross-section is drawn on the same diagram, so that they are superimposed on top of one another, at their correct height above Ordnance Datum. Once topography has been drawn in, points are

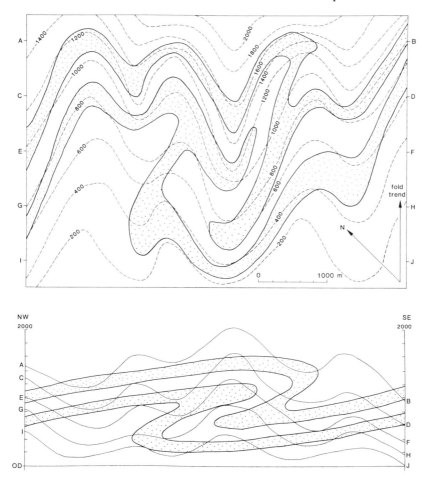

FIG. 4.12. *Above:* geological map showing the outcrop of two sandstone formations (stippled) which are affected by folding round a horizontal axis trending NE-SW. *Below:* a series of vertical cross-sections drawn across the geological map at an interval of 500 m. Note that each geological boundary can be traced across these cross-sections, so allowing a profile of the folds to be constructed.

marked off along each topographic profile, to show where the geological boundaries cross the lines of cross-section in relation to the topography. This allows the contacts between the folded layers to be drawn as smooth curves passing through corresponding points on each cross-section, so providing a fold profile in a vertical plane at right angles to the mean fold axis. It should be noted that the points where a geological boundary crosses a line of cross-section are all projected in a horizontal parallel to the trend of the mean fold axis until they lie at the same height within the plane of the fold profile.

Plunging Folds.

A different method is required to draw a profile through a series of plunging folds, since this needs to correct the distortion shown by the outcrop pattern. Consider, for example, the simple model of a circular cylinder with its axis plunging at an angle α to the horizontal, as shown in Figure 4.13. Although the profile of this cylinder is given by the circular cross-section in a plane at right angles to its axis, a horizontal cross-section through this cylinder would be elliptical. The minor axis of the ellipse gives the true diameter D of the cylinder in a horizontal direction at right angles to its plunge. This means that there is no distortion in this direction. The major axis of the ellipse is then parallel to the plunge of the cylinder, so that its length D' is given by the equation $D' = D$ cosec α. Accordingly, the distortion in this direction is such that the elliptical cross-section would need to be foreshortened in this direction by an amount equal to sin α. This would convert it into the circular cross-section which represents the true profile of the cylinder.

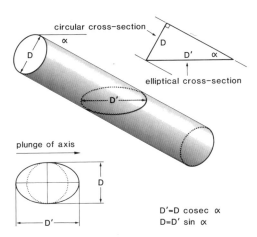

Fig. 4.13. Diagram showing that a horizontal cross-section through a circular cylinder in an ellipse with its major axis $D' = D$ cosec α where D is the diameter of the cylinder, assuming that its axis plunges at an angle α for the horizontal.

The outcrop pattern developed by a series of plunging folds in an area of low relief shows a similar distortion, which can be corrected in the same way to provide a fold profile. One method of doing this will now be described, assuming that the trend and plunge of the mean fold axis is known.

A square grid of straight lines is drawn across the geological map so that one set of grid lines is parallel to the trend of the mean fold axis, as shown in Figure 4.14. Points are marked off where the geological boundaries cross the other set of grid lines, at right angles to the trend of the mean fold axis. A rectangular grid of straight lines is then constructed on a separate

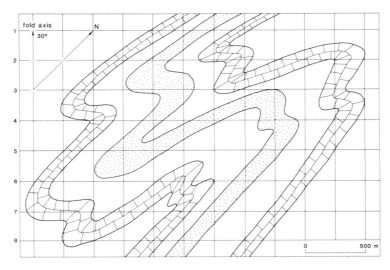

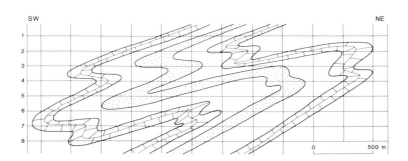

FIG. 4.14. *Above:* map showing the outcrop pattern developed by cylindrical folds plunging at 30° towards 315°N. *Below:* fold profile constructed according to the method described in the text.

diagram. One set of grid lines is drawn vertically so that they maintain the same spacing, while the other set of grid lines is drawn horizontally so that their spacing is reduced by an amount equal to sin α, in comparison with the grid lines drawn on the map. The required reduction in spacing of these grid lines may be found using trigonometrical tables, or a graphical construction may be used as shown in Figure 4.13. The points marked off on the geological map are then transferred to an equivalent position on the horizontal grid lines. Finally, the geological boundaries are drawn as smooth curves through these points, using the grid lines as a framework whereby the distorted form of the outcrop pattern can be corrected as the fold profile is sketched in.

Effect of Topographic Relief.

This construction does not take the effect of topography into account, which would further distort the nature of the outcrop pattern in areas of high relief. Consider again the simple model of a circular cylinder, plunging at an angle to the horizontal. This has an elliptical cross-section in a horizontal plane. The points lying on the circumference of this ellipse can be projected parallel to the axis of the cylinder until they lie within an irregular surface cutting across the cylinder. These points would be projected upwards by varying amounts, depending on their height above the horizontal plane of the elliptical cross-section. This means that the elliptical cross-section of the cylinder becomes distorted, wherever this cross-section is not formed by a plane. The outcrop patterns developed by folded rocks in areas of high relief show a similar distortion, as shown in Figure 4.15, which can

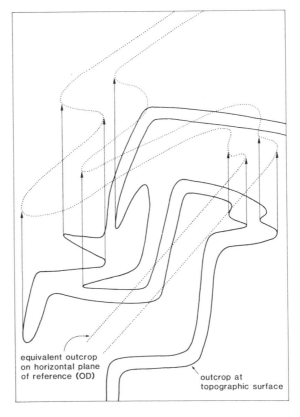

equivalent outcrop
on horizontal plane
of reference (OD)

outcrop at
topographic surface

FIG. 4.15 Effect of topography on the outcrop pattern of folded rocks, showing how the geological boundaries need to be projected down-plunge to give the true form of the geological structure. Solid lines: geological boundaries corresponding to the topography shown in Figure 4.16. Dashed lines: geological boundaries on a horizontal plane at Ordnance Datum.

only be corrected if the height of the topographic surface is taken into account, relative to the plunge of the mean fold axis. The construction now to be described provides this correction.

A straight line *AB* is first drawn at right angles to the trend of the mean fold axis, along the down-plunge side of the geological map (see Fig. 4.16).

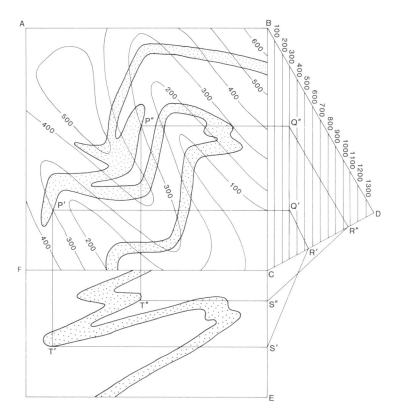

FIG. 4.16. Construction of a fold-profile for an area where the outcrop pattern is affected by the topography. See text for details.

A second line *BC* is drawn perpendicular to this line, so that it is parallel to the trend of the mean fold axis, along one side of the map. A right-angled triangle *BCD* is then constructed on the line *BC* as its hypotenuse, so that the angle *CBD* is equal to the plunge of the mean fold axis. The triangle is subdivided by a series of elevation lines, drawn parallel to the hypotenuse *BC*, and corresponding in height to the topographic contours shown on the geological map. This triangle then provides a vertical cross-section parallel to the trend of the mean fold axis. This means that points can be projected within this plane parallel to the plunge of the mean fold axis, as given by the line *BD*. Finally, a line *CE* is drawn as a continuation of the line *BC*, while the line *CF* is drawn at right angles to this line, along the up-plunge side of the geological map.

A point *P* is then chosen on the geological map where a boundary crosses a topographic contour. Two lines *PQ* and *PT* are drawn through this point, parallel to *AB* and *BC*, respectively. The line parallel to *AB* is continued until it meets the elevation line at the point *Q*, corresponding to the height of the topographic contour. This point of intersection is then projected parallel to *BD* until it meets the line *CD*, in the point *R*. It is then transferred, using a pair of dividers, to the point *S* on the line *CE* so that *RD* = *SE*. Finally, it is projected parallel to the line *CF*, until it meets the line drawn through the original point parallel to *BC* in the point *T*. This point is marked on the plane of the fold profile. This procedure is repeated for each point where the geological boundary crosses a topographic contour. There should then be sufficient points on the plane of the fold profile to draw in the form of the geological boundary as a smooth curve passing through these points. The other boundaries shown on the geological map are transferred in the same way, in order to complete the fold profile.

Construction of the Axial Trace.

It has already been stated that the axial trace of a fold represents the outcrop of its axial plane at the earth's surface, while the axial plane is defined by a surface passing through a successive fold hinge along its closure. Each fold hinge is represented by a point of maximum curvature in the profile of a surface affected by the fold. However, if the fold hinge plunges at an oblique angle within an inclined axial plane, this fold hinge does not necessarily pass through the point of maximum curvature, as shown by the geological boundary representing the outcrop of this surface (see Fig. 4.7). The axial trace cannot then be drawn as a line passing through all the points of maximum curvature along the outcrop of a particular fold. This means that a separate method is required to construct its axial trace.

If a fold profile has already been prepared for the area under consideration, it is a simple matter to draw in the axial plane of a particular fold on the fold profile, passing through all the points of maximum curvature along its closure. The method of construction, whereby the fold profile was prepared in the first place, is then reversed so that points lying along the axial plane are transferred to a corresponding position on the geological map. A line drawn through these points, taking the effect of topography into account, then represents the axial trace of the fold.

Parallel Folds in Cross-Section

The profile of a cylindroidal fold can only be constructed if the outcrop pattern shows the form of the fold closure on the ground. Unless the topography has an appreciable relief, this requires the fold to plunge at a considerable angle. The morphology of the fold can then be described objectively. However, the profile of a horizontal fold is difficult to construct from the outcrop pattern in an area of low relief, simply because such a fold rarely closes along the ground under these circumstances. Unfortunately, this is perhaps the most common of all the circumstances under which a fold profile needs to be drawn in order to show the geological structure of an area.

Assumption of Parallel Folding.

If such an area consists of sedimentary rocks which are not affected to any degree by regional metamorphism, it is commonly assumed that the folding was accomplished as a result of bedding-plane slip. Since this tends to preserve the orthogonal thickness of the individual beds, it is likely that parallel Class 1B folds would tend to develop under these conditions. This assumption is supported by observing that the small-scale folds seen in the field often have the form of parallel folds. Naturally, the vertical cross-section drawn across such an area only show the geological structure which would exist if this assumption were correct. However, it should be admitted that geological maps are commonly accompanied by vertical cross-sections which have been prepared according to this assumption, so that they are essentially subjective.

Involutes and Evolutes.

The geometry of parallel folds can be described as a series of parallel curves, generated as the involutes of a fixed curve known as an evolute. Consider, for example, a straight line *ABC* rolling about a fixed curve so that no sliding takes place at the point of contact (see Fig. 4.17). This straight line is always a tangent of the evolute. A point *C* on this straight line traces out a second curve to from an involute of the fixed curve. Evidently, there are an infinite number of involutes forming a single family in that they are all related to a single evolute. This means that any two involutes can be chosen to form the bounding surfaces of a layer affected by parallel folding.

A circle is a particular type of involute, generated by a point on a straight line moving so that it remains at a constant distance from a fixed point. This straight line forms the radius of the circle, so that the straight line rolling about the evolute in the general case can be termed the radius of the involute. The evolute of a circle is a fixed curve which has degenerated to a single point at its centre.

Although it is a special case, the circle has certain properties which are typical of involutes in general, as shown in Figure 4.18. Firstly, the tangent

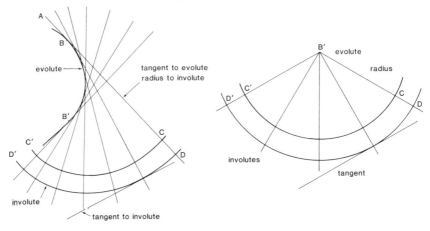

FIG. 4.17 Diagram showing how a pair of parallel curves can be considered to form as involutes to an given evolute. *Left:* general case. *Right:* circular arcs as a special case.

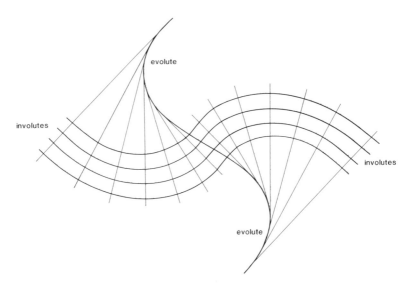

FIG. 4.18. Diagram illustrating the geometrical features of parallel folding as expressed in terms of involutes and evolutes.

to a circle is always perpendicular to its radius at the same point. By analogy, any line drawn as the tangent to an involute is perpendicular to the radius of the involute at the same point. This means that the radius of an involute can be found simply by drawing a line perpendicular to its tangent, so that it forms a normal to the involute at the point of tangency. Secondly, the radius of a circle always passes through its centre. It has already been mentioned that the radius of an involute is tangential to its evolute, which is the equivalent relationship expressed in more general terms. Accordingly, the evolute forms an envelope to all the lines which can be drawn as normals to the involute. Thirdly, a circle has a radius of curvature which corresponds in length to its radius. Likewise, the radius of curvature shown by an involute at a particular point is given by the length of the radius at that point, as measured from the involute to its point of contact with the evolute. This means that any two lines drawn as radii of the same involute differ in length by an amount equal to the distance between their points of contact, as measured along an arc of the evolute between these points.

The latter relationship implies that involutes belonging to the same family always differ in radius by an amount equal to the orthogonal thickness of the intervening layer. Since the orthogonal thickness of this layer must therefore remain the same, it is self-evident that the surfaces in a parallel fold form a family of involutes which are all related to the same evolute. Accordingly, any line drawn as the radius of a particular involute intersects all the other involutes at right angles, so that this line forms a common radius to the whole family. These relationships form the basis for the two methods now to be described for the construction of a vertical cross-section through a parallel fold with a horizontal hinge.

Tangent-Arc Method.

Unless it is a circular arc, an involute generally varies in curvature along its length. Consider, for example, two points A and B lying on the same involute, as shown in Figure 4.19. If the normals to the involute are drawn through these points, they are found to be tangential to the evolute at C and

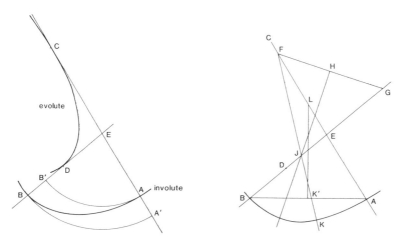

Fig. 4.19. A: Diagram showing how the mean radius of curvature can be determined for the arc of a curve. B: Construction to interpolate an intermediate dip at K'. See text for details.

D. The length of the normals *AC* and *BD* then gives the radius of curvature for the involute at the points *A* and *B*, respectively. Since the form of the evolute is generally not known, it would have to be constructed as an envelope to all the lines which can be drawn as normals to the involute. This would allow the precise radius of curvature to be determined at a given point on the involute, corresponding to a particular centre of curvature on the evolute.

This construction can be simplified by considering how the lines *AC* and *BD* drawn as normals to the involute intersect one another at the point *E*. This point gives a mean centre of curvature for the involute along the arc *AB*. The circular arcs *AB'* and *A'B* can then be drawn about the mean centre of curvature *E*, so that they pass through *A* and *B*, respectively. If the normals *AE* and *BE* are equal in length these arcs would coincide with one another to form the arc *AB* of the involute, as required. This would mean that the involute had a constant radius of curvature along the arc *AB*. Otherwise, there would be a discrepancy between the circular arcs *AB'* and *A'B*. However, this discrepancy is likely to be slight, unless there is a marked change in the radius of curvature along the arc *AB*. If so, this arc can only be constructed satisfactorily by dividing it into a number of short segments, each of which can be treated as a circular arc. This implies that the attitude of the involute is known at a number of intermediate points along its length, which may not be the case. The construction now to be described avoids this difficulty.

It is possible to interpolate an intermediate dip between the two points *A* and *B* where the attitude of the involute is known, which allows the involute to be constructed as a single curve between these points, as shown in Figure 4.19. As previously, the lines *AC* and *BD* are drawn as normals to the bedding at *A* and *B*, so that they intersect one another at the point *E*. The construction should be used if it is found that the circular arcs *AB'* and *A'B* drawn with their centre at *E* to pass through *A* and *B*, respectively, do not coincide with one another. The normals *AC* and *BD* should first be extended well beyond their point of intersection *E*. Assuming that the dip is less at *A*, select an arbitary point *F* on the normal *AC* drawn through *A*, so that it lies farther away from the point *L* in which the perpendicular bisector of the line *AB* intersects the normal *AC*. Mark off the point *G* on the normal *BD* through *B*, so that the lines *AF* and *BG* are equal in length. Joining the points *F* and *G* with a straight line, construct the perpendicular bisector *HJ* of this line so that it intersects the normal *BD* drawn through *B* in the point *J*. The points *F* and *J* are then the two centres of curvature which allow the involute to be drawn as a single curve between *A* and *B*, even although one of these centres has been chosen in an arbitrary fashion. This is done by drawing a circular arc *AK* though the point *A*, with its centre at *F*, so that it meets the line *FJ* in the point *K*. The interpolated dip of the bedding is at *K'* perpendicular to this line. The curve *AK* is then continued by drawing the circular arc *KB* with its centre at *J*, so that it passes through the point *B*. This must be the case since *AF* = *BG* = *FK* while *GJ* = *FJ*, so that *JK* = *JB*. Although the curve *AKB* has been constructed according to an arbitary assumption, it is a close approximation to the true form of the involute between *A* and *B*.

Construction using the Tangent-Arc Method.

The details of the tangent-arc method can now be given as they apply to the construction of a vertical cross-section through a parallel fold. A topographic cross-section is first drawn along a suitable line at right angles to the mean fold axis, as shown in Figure 4.20. Since this is horizontal, the line of cross-section is perpendicular to the strike of the beds. Wherever the structural observations shown on the geological map concern the dip of the bedding, they should be transferred to their appropriate position on the cross section. If necessary, these observations can be projected parallel to the strike of the beds wherever they occur at a short distance on either side of the cross-section. Mark off the points where the geological boundaries cross the line of the cross-section.

Normals are then drawn perpendicular to the dip of the bedding at each locality *A, B, C, D* and so on, as shown in Figure 4.20. Adjacent lines are

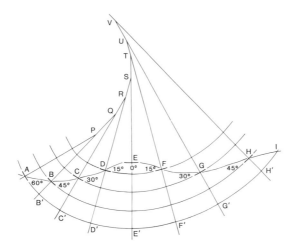

FIG. 4.20. Diagram showing how the tangent arc method can be used to draw a vertical cross-section through a parallel fold, given the dip of the bedding as shown. See text for details.

projected upwards or downwards in pairs so that they intersect one another in the points *P, Q, R* and so on. These points would only coincide with one another if the bedding was folded with a constant radius of curvature to form a concentric fold in the strict sense. Now identify a particular horizon which is repeated on the fold limbs midway between the axial traces of the folds. Supposing that the horizon outcrops at A, draw the circular arc AB', with its centre at P, passing through A to meet the normal drawn through B in the point B'. This curve is continued by drawing the circular arc $B'C'$, with its centre at Q, from B' until it meets the normal drawn through C in the point C'. Repeat this procedure until the curve so constructed intersects the topography once more.

This point of intersection may or may not correspond to the actual out-crop of this horizon on the opposite fold limb. If the correspondence is reasonably close, it may be assumed that the curve so constructed is a close approximation to the form of the folded horizon between the two points where it outcrops on the opposite limbs of a fold. Any slight discrepency can be adjusted by freehand since the method of construction tends to in-troduce drafting errors in any case. These mainly arise because it is difficult to locate precisely the point of intersection where two lines cross one another at a low angle. Further errors may be introduced wherever the form of the topographic surface is uncertain, simply because a topographic con-tour does not pass through a point of structural observation. It should also be remembered that bedding tends to show local variations in dip, which are not necesarily related to the form of the regional structures.

If there is a marked discrepancy between the actual outcrop of the folded horizon, and its predicted position on the opposite fold limb, it may be possible to correct this discrepancy by interpolating an intermediate dip according to the method already described. This is best done wherever structural control is lacking along the line of cross-section. For example, an intermediate dip could be interpolated between the two points which show the greatest difference in dip, particularly if these points are far apart. However, if there is an adequate number of structural observations, evenly spaced along the line of cross-section, such a discrepancy becomes difficult to correct. It might then be concluded that the assumption of parallel folding, which forms the basis for this method of construction, cannot be maintained in the case under consideration.

If such discrepancies in outcrop can be corrected as they arise, the method of construction can be repeated until the folded horizon has been traced throughout the cross-section. It is then possible to draw in all the other horizons which outcrop along the line of cross-section, using the same method, in order to complete the vertical cross-section.

It should be noted that a folded horizon should be drawn as a straight line wherever the bedding does not change in dip along the line of cross-section. This means that the lines drawn as normals to the bedding at these points of equal dip would be parallel to one another, so that they only intersect one another at infinity. However, a difficulty is encountered wherever there is only a slight difference in dip between adjacent points, since the normals to bedding at these points would only intersect one another at a considerable distance. There is a construction which can be used to draw in an arc of the folded surface between these normals, as shown in Figure 4.21. Assume that

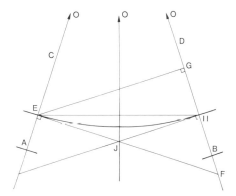

FIG. 4.21 Construction for drawing in a curve where there is only a slight difference in dip between adjacent points. See text for details.

the lines *AC* and *BD* are the normals to the bedding at *A* and *B*, respectively. If the arc needs to be drawn through the point *E* on the normal *AC*, construct the lines *EF* and *EG* so that they are perpendicular to *AC* and *BD*, respectively. Bisect the angle *FEG* so that the bisectrix *EH* intersects the normal *BD* in the point *H*. Draw in the line *HJ* as the perpendicular to *BD* so that it intersects *EF* at the point *J*. The arc then passes through the point *H*, simply because the triangles *EOJ* and *HOJ* are congruent to one another, so that *EO* = *OH*, as required. The arc *EH* can be constructed by freehand, using the tangents *EJ* and *GJ* as guide-lines.

Layer-Thickness Method.

This construction is based on a particular property of involutes which all belong to the same family. If *t* is the orthogonal thickness of the layer lying between two such involutes, a circle of radius *t* can always be drawn at any point on one involute so that it is tangential to the other involute, as shown in Figure 4.22. This means that an involute can be constructed as an envelope to all the circles which can be drawn with their centres on any other involute belonging to the same family, provided that these circles are always equal in radius to the orthogonal thickness of the intervening layers. The geological boundaries shown on a geological map correspond in position to a series of points on a family of involutes, while the stratigraphic thickness of the intervening formations gives the orthogonal thickness of the layers lying between these involutes. The latter information is often

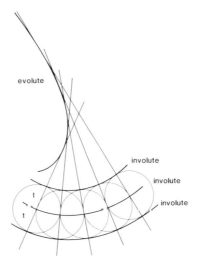

FIG. 4.22. Diagram illustrating how a series of circles can be drawn on one involute so that they are tangential to two other involutes of the same family.

given by the stratigraphic column which accompanies a geological map, if it is drawn to scale.

The construction first requires a topographic cross-section to be prepared along a suitable line on the geological map, at right angles to the mean fold axis. The position of the geological boundaries should be shown on this cross section, corresponding to the localities *A, B, C, D* and so on, as shown in Figure 4.23. The stratigraphic formations *P, Q* and *R* then outcrop as *AB, BC, CD* and so on. Let the orthogonal thicknesses of these formations be *p, q* and *r,* respectively, Now select a particular horizon which outcrops on the fold limbs midway between the axial traces of the folds. Taking each boundary in turn, a circular arc should be drawn with its centre at this point, equal in radius to the orthogonal thickness of the stratigraphic for-

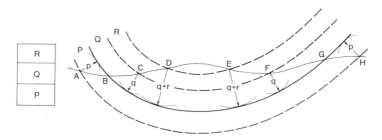

FIG. 4.23. Diagram showing how the layer-thickness method can be used to draw a vertical cross-section through a parallel fold, given the stratigraphic thickness of the formations as shown. See text for details.

mations intervening between this boundary and the horizon so selected. Thus, if this horizon passes through *B,* circular arcs should be drawn at *A* with a radius *p,* at *C* with a radius *q,* at *D* with a radius *q + r,* and so on. This can be done for all the points along the line of cross-section, whether the selected horizon occurs above or below the topographic surface. This horizon can then be constructed as a smooth curve tangential to all these circular arcs so that it passes through its boundaries as shown on the line of cross-section. This procedure can be repeated for all the other horizons outcropping along the line of cross-section, in order to complete the vertical cross-section.

This method does not provide any independent check that the basic assumption of parallel folding has correctly applied in the case under consideration, except that the width of outcrop must always be greater than the orthogonal thickness of the corresponding formation. Obviously, it requires the orthogonal thickness of the stratigraphic formations to be known. This can be measured in the field, horizontally along a line of section or vertically down a bore-hole or mine shaft, provided that the dip of the bedding is taken into account. However, it can also be determined from the dip of the upper and lower contacts to a formation, where this is folded, as shown in Figure 4.24. Consider the lines *AC* and *BC* drawn as

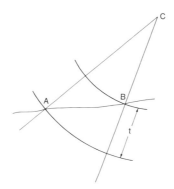

Fig. 4.24. Diagram showing how the orthogonal thickness of a folded layer can be determined.

normals to the bedding at *A* and *B,* so that they intersect one another at the point *C.* The difference in the radii of curvature *AC - BC* then gives the orthogonal thickness *t* of the intervening formation, assuming that the contacts are folded in a concentric manner.

Balanced Cross-sections through Parallel Folds.

Even if a vertical cross-section can be constructed through a parallel fold using these methods, it must satisfy an important condition in order to provide a realistic model of the geological structure. Thus, it is assumed that the orthogonal thickness of a layer remains the same as a result of parallel folding. The mechanism of bedding-plane slip does not allow any extension to occur parallel to the fold hinges, while the density of the folded rocks is unlikely to change by a significant amount. This means that the length of the layer must remain the same in the plane of the cross-section, so that its volume can be conserved. Accordingly, all the surfaces affected by parallel folding must have the same length, as measured between those areas where no bedding-plane slip has taken place. If this condition is satisfied, the result is known as a balanced cross-section. It presents an interpretation of the geological structure which is internally consistent, at least.

Most cross-sections through parallel folds are only balanced within certain limits, even if they have been constructed using the methods already described in this section. This may be illustrated by taking concentric folds as a special case. Figure 4.25 shows an antiform flanked on either side by a

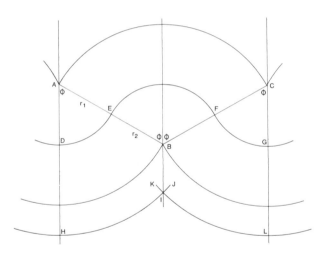

Fig. 4.25. Diagram illustrating the limits of a balanced cross-section through a concentric fold. Note that the arcs HJ + KL are equal in length to the folded surface DEFG.

synform, so drawn that the fold limbs dip at an angle ϕ. It is assumed that no slip has occurred on the bedding planes where they are horizontal in the troughs of the synforms. Any folded surface *DEFG* lying between the centres of curvature *A, B* and *C* must have a length equal to $2 (r_1 + r_2) \phi$, where r_1 and r_2 are the radii of curvature for the folds as shown. Since $r_1 + r_2$ is a constant, equal in length to *AB* or *BC*, any folded surface lying between the centres of curvature *A, B* and *C* must have the same length. This means that the cross-section is balanced within these limits.

The folded surfaces forming the limits of a balanced cross-section form cusps as they pass through the centres of curvature *A, B* and *C,* where the radius of curvature is zero. It has been the convention to extend the cross-section beyond these limits simply by rounding off these cusps, wherever they are developed by the overlap of folded surfaces which lie beyond the centres of curvature. However, the cross-section cannot then be balanced, since these surfaces are effectively shortened by this method. Thus, it can be seen from Figure 4.25 that the folded surfaces *HIL* lying below the centre of curvature *B* for the antiform would gradually become shorter as they approach the form of a straight line at depth. Since it has already been argued that parallel folding tends to preserve the original length of the folded surfaces, this means that these surfaces are too short wherever they occur beyond the limits of a balanced cross-section. Other features must, therefore, be introduced into the cross-section in order to preserve the original lengths of the folded layers. This can be done in several ways.

Zones of Contact Strain.

The buckling of layers under compression tends to produce folds which are approximately parallel in style. This buckling typically affects the competent layers, such as sandstones and limestone, which are more resistant to the deformation. It allows the incompetent layers, such as shales and evaporites, which are less resistant to the deformation, to flow into the spaces formed by these buckles. The nature of the folding then depends on the relative spacing of the competent and incompetent layers.

If a single layer is surrounded on either side by incompetent material, it will buckle into a series of quasi-parallel folds with a particular wavelength. However, these buckles gradually disappear as they are traced into the surrounding material, as shown in Figure 4.26. This occurs over a distance which is approximately the same as the intial wavelength of the folds, so

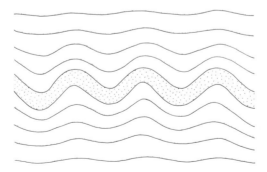

FIG. 4.26. Diagram showing zones of contact strain adjacent to a competent layer (stippled) which is affected by a series of quasi-parallel folds.

defining a zone of contact strain around the buckled layer. Beyond this zone of contact strain, the incompetent material has simply been affected by a uniform compression parallel to the buckled layer, without any folding taking place. However, folds are formed within this zone of contact strain, which belong to Class 1A or Class 3, according to whether they are developed on the inner or outer arcs of the buckled layer. Thus, the buckles die out upwards as Class 1A folds and downwards as Class 3 folds, or vice versa, according to whether they are antiforms or synforms, respectively.

Folds of this type are commonly developed by pegmatite veins in high-grade metamorphic rocks, such as gneisses and migmatites, where they may have the form of elasticas. They are known as ptygmatitic folds (see Fig. 4.27). However, sedimentary rocks are also affected by folding of this type,

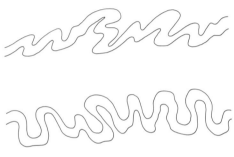

FIG. 4.27 Diagram showing characteristic form of ptygmatic folds as developed by pegmatite veins in high-grade gneisses and migmatites.

wherever a single competent layer is isolated in a highly incompetent matrix, such as rock salt, anhydrite or gypsum. The stratified nature of sedimentary rocks means that such a competent horizon is likely to be repeated within the sequence. These layers tend to fold independently of one another if they are separated by a sufficient thickness of incompetent material. This means that the geological structure of an underlying layer at depth is difficult if not impossible to determine from the evidence presented by the folding of a competent layer exposed at the earth's surface, as can be seen from Figure 4.28. The deformation would then have a very complex pattern where it affected the incompetent material lying between these layers.

If the competent layers are separated by lesser amounts of incompetent material, the buckles formed in these layers tend to interact with one another, so that a single set of folds would be formed, affecting all the layers. These folds have a form typical of composite similar folds, in that the competent layers tend to buckle into Class 1B parallel folds while the incompetent layers thicken into the fold hinges for form Class 3 folds, as shown in Figure 3.19. Such folds commonly have wide and well-rounded hinges. However, as the layers of incompetent material become thinner, in

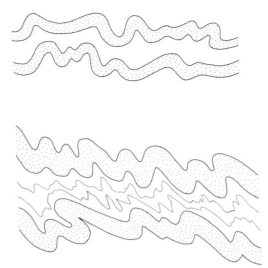

Fɪɢ. 4.28. Disharmonic folding of competent layers (stippled) separated from one another by appreciable thickness of very incompetent material.

comparison with the thickness of the competent layers, the hinge zones become narrower and more angular. This leads to the development of chevron folds, with straight limbs separated from one another by angular hinges. If the incompetent layers are very thin, there is generally insufficient material to thicken into the fold hinges to form Class 3 folds, as required. This may allow cavities to open at the hinge zone between the competent layers, which are filled with vein material to form saddle reefs. Alternatively, it may cause the buckling to affect the whole sequence of different layers as a single unit, known as a multilayer. This would then buckle to form folds on a larger scale.

Décollement.

The buckling of a competent layers can also be accommodated if these layers become detached from the underlying substratum along an incompetent horizon, as shown in Figure 4.29. This process is known as décollement, from the French verb to become unstuck. This term has now passed into English, in common with several other words applied to tectonic or structural features. Such a décollement generally occurs along one or more bedding-plane faults, which follow a particularly incompetent horizon below the competent layers. This horizon is commonly formed by shales which are poorly consolidated or by evaporite deposits such as rock salt, anhydrite or gypsum. It appears that décollement is more likely to occur if such an incompetent horizon lies near the base of a sedimentary sequence,

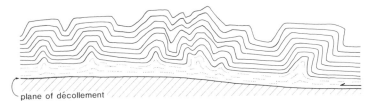

Fig. 4.29. Development of a plane of décollement below a folded layer.

above a rigid basement of igneous and metamorphic rocks. This means that the folding only affects the sedimentary rocks acting as a cover to the older basement, giving rise to folds known as "plis de couverture". They may be compared with folds, termed "plis de fond", which would affect the basement rocks as well as the sedimentary cover.

It is possible to determine the depth of folding as a measure of the vertical thickness of the rocks lying above the décollement, as shown in Figure 4.30. This is done by selecting a particular surface affected by the folding, as shown in a vertical cross-section. Two points *A* and *B* on this horizon are then located in an equivalent position at either end of the cross-section. The straight line *AB* is drawn through these points, and its length measured. The length *A′B′* of the folded horizon between these points can also be measured, using a curvimeter. The difference in length *A′B′ - AB* gives the amount of shortening which has taken place. However, this shortening must be counter-balanced by an increase in the vertical thickness of the rocks lying above the décollement, which occurs as a result of the folding. This increase must be sufficient to preserve the area of the rocks lying between the décollement and the folded horizon *A′B′* in the plane of the cross-section. The area lying between the straight line *AB* and the folded horizon *A′B′* should therefore be measured with a planimeter. Note that any areas lying above the straight line *AB* should be considered positive, while any areas lying below this line should be subtracted from the total as negative quantities. As shown in Figure 4.30, the area so measured must be

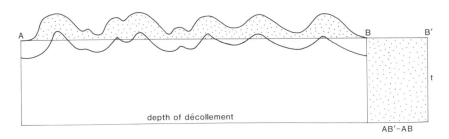

Fig. 4.30. Diagram showing how to estimate the depth of folding above a plane of décollement. See text for further details.

equal to the area $\Delta = t(A'B' - AB)$ at the side of the cross-section, where t is the vertical depth to the decollement, if it is assumed that A and B are still at the same height above the plane of décollement. This means that the depth of folding can be found by dividing the area below the folded horizon by the lateral shortening $A'B' - AB$.

Changes in Fold Morphology.

The buckling of competent layers above a décollement leads to a lack of space within the fold cores, which generally becomes more acute as the folding proceeds. The cause has already been mentioned in that the folded layers cannot simply preserve their original lengths beyond the limits of a balanced cross-section, as defined by the centres of curvature for the individual folds. However, these folds are produced initially with a radius of curvature, which corresponds to a centre of curvature at infinity. This means that the centre of curvature in a concentric fold, for example, gradually approaches the folded layers during the course of the folding, as the radius of curvature decreases from infinity to a finite value. In other words, the cross-section is only balanced within certain limits, which converge on one another as the centres of curvature move towards the folded layers. The material lying beyond these limits must deform in order to accommodate the folding as it continues. This can be achieved if this material is sufficiently plastic to be squeezed from the fold cores. Commonly, this material becomes folded in a complex manner as it is expelled in this way, as shown in Figure 4.31C. This mechanism can only work if the material underlying the competent layers is relatively incompetent. However, increasing amounts of incompetent material need to be expelled as the centre of curvature for a particular fold moves towards the competent layers which control the folding. As the centre of curvature passes through a particular layer, this can no longer be affected by parallel folding in conformity with these competent layers, even if this was previously the case. This layer may simply contract, if it is relatively incompetent, or it may buckle into a series of small-scale folds which can accommodate the lateral shortening within the fold-core, if it is relatively competent.

The expulsion of material from the core of a parallel fold may encounter some resistance if this material is difficult to deform. Such a resistance to deformation may so modify the shape of the developing fold that it becomes a box fold. This has the effect of widening the fold core in such a way that there is no longer any lack of space as the fold develops. Alternatively, thrust faults may be developed on the fold limbs, as shown in Figure 4.31A. Such faults are likely to form once the centre of curvature for a particular fold has crossed into the competent layers which originally controlled its development. The lack of space is then solved by the outward

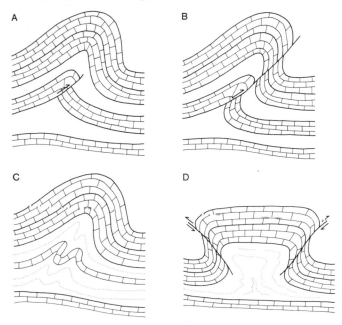

FIG. 4.31. Development of thrusts and associated structures as a result of parallel folding.

thrusting of the fold core over the rocks forming the fold limbs. As shown in Figure 4.25, this mechanism preserves the original lengths of the competent layers lying within the fold core.

Such thrusts may be associated with the development of box folds, in which case a thrust fault is present on both fold limbs, dipping towards the core of the fold, as shown in Figure 4.31D. The symmetrical development of these thrust faults accentuates the box-like form of the fold. Alternatively, only a single thrust fault may be present, situated on the steeper limb of the fold. The asymmetrical development of such a thrust fault leads to the formation of what is generally known as an overthrust antiform, as shown in Figure 4.31B. It may well be assumed in most cases that this thrust fault would penetrate downwards until it reached the underlying plane of décollement, if such is present at depth. It can be noted that low-angle overthrusts commonly end in a series of thrust faults which steepen towards the surface, where they are associated with the development of antiformal folds. Once such an overthrust antiform is produced, the thrust plane may act as a plane of weakness, along which incompetent material can be expelled from the fold core. This material is ususally formed by evaporite deposits such as rock salt, anhydrite or gypsum, even although shales may also be affected if they are poorly consolidated. The upward expulsion of this material gives rise to piercement folds or diapirs.

CHAPTER 5

Joints, Veins and Faults

Introduction

THIS chapter deals with the various structures that are formed by the brittle fracturing of rocks under stress. This process of mechanical failure results in the formation of fractures which disrupt the original continuity of the rock. They can be divided into extension fractures and shear fractures, according to their mode of origin.

Extension fractures are formed if the wallrocks on either side of the fracture have moved apart, so that a displacement has taken place *across* the fracture. They are produced by mechanical failure under tension. However, it should be realised that absolute tension is rarely developed at any depth in the earth's crust, due to the weight of the overlying rocks. Instead, the hydrostatic pressure exerted by a pore fluid can convert a compression into an effective tension. This means that extension fractures tend to form at right angles to a direction of minimum compression P_3, as shown in Figure 5.1.

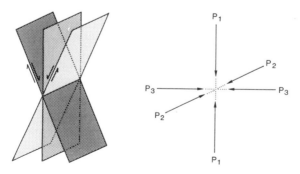

FIG. 5.1. Configuration of stresses developed within a body as the result of applied forces. P_1, P_2 and P_3 are the maximum, intermediate and minimum compressions acting at right angles to one another. The vertical plane on the left represents an extension fracture, while the other two planes correspond to shear fractures.

Shear fractures are formed if the wallrocks on either side of the fracture have moved so that a displacement has taken place *along* the fracture. They are produced by mechanical failure under shear. Theroretical considerations combined with experimental evidence indicate that shear fractures form as conjugate sets, lying at an acute angle of approximately 30° on either side of a direction of maximum compression P_1. The displacement on these fractures is directed towards the apex of this acute angle, as shown in Figure 5.1.

Since the directions of maximum and minimum compression occur at right angles to one another, it is found tha a single set of extension fractures bisects the acute angle between the conjugate sets of shear fractures as shown in Figure 5.1 These fractures all intersect one another in a direction of intermediate compression P_2, which lies at right angles to the plane containing the directions of maximum and minimum compression.

The structures formed by the brittle fracturing of rocks are known variously as joints, fissures and faults. Joints are formed if there is no visible displacement of the wallrocks on either side of the fracture. However, some movement must have occurred for the joint to form. Joints are classified as extension joints or shear joints according to the nature of this movement. Fissures are formed if the wallrocks on either side have moved apart. Since empty fissures cannot exist at any depth, they are generally represented by igneous intrusions such as dykes, sills and sheets, and by mineral veins. Faults are formed if the wallrocks on either side have been affected by displacements along the fracture. They are the most important type of fracture as far as the geological structure of the earth's crust is concerned.

Joints and Their Pattern

Joints are the clean-cut fractures which can be seen in virtually every exposure at the earth's surface. They are the commonest of all the structures developed in a rock after its formation. They exert an important influence on the topography, since they often control the form of a coastline or the nature of a drainage system. However, joints are rarely shown on a geological map, unless it is a large-scale plan produced for the purposes of mining or engineering geology.

Joints typically occur as parallel fractures cutting the rock at a regular interval, which varies from a few millimetres to many metres, as shown in Figure 5.2. Such fractures are known as systematic joints. They are termed master joints if they have a greater than average extent. Their planar nature serves as a contrast to the irregular, curved or conchoidal form of non-systematic joints.

Joint Sets and Joint Systems.

Systematic joints are found as joint sets, formed by all those fractures which are virtually parallel to one another. A joint system is produced if

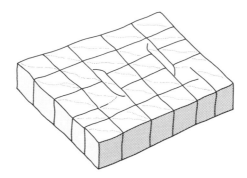

FIG. 5.2. Typical pattern developed by systematic joints.

more than one joint set is present. The joints belonging to different sets intersect one another at an angle known as the dihedral angle between the joint sets.

The character of a joint system depends on the relative development and arrangement of the various joint sets. Each joint set may be equally well developed, forming a series of joints which cut across one another without interruption. Alternatively, a particular joint set may form the dominant structure in the rock, giving rise to master joints against which the other joints terminate. The joints lying between the master joints often occur in the form of curved or irregular fractures. They are known as cross joints if they are approximately at right angles to the master joints.

Our understanding of joints and their formation is hindered by two factors. First, it is difficult to distinguish extension joints from shear joints on the basis of displacement simply because joints are defined as fractures which show no visible displacement. Second, it is difficult to determine the relative age of joints, since they have little or no effect on one another. This means that it is often impossible to determine which joint sets are contemporaneous, and so related to one another. It is usually necessary to consider their relationships to other structures in order to determine their origin.

Jointing as a Superficial Phenomenon.

Some joints are clearly superficial, since they are found to disappear at depth. For example, massive rocks such as granites and some sandstones are commonly affected by sheet joints, giving rise to a structure known as sheeting, as shown in Figure 5.3A. These joints form curved fractures which tend to occur parallel to the topographic surface, except in areas of rapid erosion. Although they may be only a few centimetres apart at the surface, this spacing increases with depth until they eventually disappear. These fractures are entirely independent of all the other structure in the rock. It

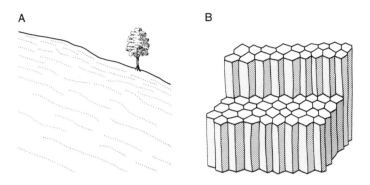

FIG. 5.3. Joint patterns in igneous rocks: A: Sheeting. B: Columnar jointing.

has been suggested that they form in response to the residual stresses which develop during the uplift and erosion of deep-seated rocks. If so, they would have the form of extension joints lying at right angles to the direction lying of maximum relief.

Jointing in Igneous Rocks.

Other joints are formed during the emplacement and consolidation of igneous rocks. For example, the outer parts of an igneous intrusion may be completely consolidated, while the interior is still fluid. Continued intrusion of magma under pressure would then exert stresses on the consolidated rocks forming the margins of the intrusion. These stresses may be sufficient for fracturing to occur. The fractures so formed often extend outwards into the country rocks around the intrusion, while they may be intruded by magma to form minor intrusions such as dykes, sills and sheets. The resulting fracture pattern usually shows a close relationship to the flow structures formed during the earlier stages in the emplacement of the intrusion. Fracture patterns of this sort are commonly developed in granite plutons.

Jointing may also develop in response to the contraction which occurs during the cooling of igneous rocks. This commonly results in a system of extension joints which divides the rock into a series of pentagonal or hexagonal columns, as shown in Figure 5.3B. This structure is known as columnar jointing. It is found in lava-flows and minor intrusions. The columns are formed perpendicular to the cooling surface, so that they are often vertical.

Jointing in Sedimentary Rocks.

Although these processes are important, joints are best developed in flat-lying or folded sediments, where they occur as well-defined sets of systematic joints at right angles to the bedding. Such systems in flat-lying sediments are commonly formed by two joint sets at right angles to one another. These joints are most likely to be extension fractures. Another two joint sets may be developed at an acute angle of approximately 60° to one another. These joints are most likely to be shear fractures. The resulting joint system is symmetrical since the angles between the shear joints are usually bisected by the extension joints. All these joints are vertical if the bedding is horizontal.

A similar pattern is also developed in folded rocks, as shown in Figure 5.4, where one joint set lies at right angles to the fold axis, while another set is developed parallel to this direction, so that they are arranged at right angles to one another. The first set consists of cross joints, formed parallel to the plane of the fold profile. The other set is formed by longitudinal joints. Both sets are likely to be formed by extension fractures. They can only be described as dip joints and strike joints if the fold axis is horizontal.

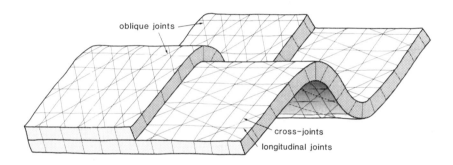

Fig. 5.4. Joint patterns in folded rocks.

Longitudinal joints occasionally form as conjugate sets of shear fractures. They generally occur in the vicinity of the fold hinge, where they are inclined at a high angle to the bedding. More commonly, conjugate sets of shear fractures are developed as oblique joints at right angles to the bedding. These joints intersect one another in an acute angle which is usually bisected by the cross joints. They would be produced in response to a direction of maximum compression at right angles to the fold axis. However, these oblique joints are sometimes arranged so that the acute angle between them is bisected by the longitudinal joints.

Similar joint patterns are developed in metamorphic rocks which have undergone folding and deformation. The cross joints in this case are often found at right angles to a dominant linear structure in the rock. They are commonly occupied by quartz veins.

The various sets of extension joints and shear joints which result in the joint systems just described are closely related to one another. More complex patterns are developed if the joint sets differ in age. Such patterns are commonly found in igneous and metamorphic rocks, and in sedimentary rocks which have been affected by a long history of earth movements.

Fissures and Mineral Veins

Mineral veins simply represent the infilling of fissures which have opened in the surrounding country-rocks. The minerals in such veins are deposited from aqueous solutions which flowed through the fissure as it opened. These solutions often react with the country-rock, particularly if it is limestone, giving rise to irregular masses of alteration and replacement products similar to the minerals forming the vein itself. These minerals include gangue minerals such as quartz, calcite, dolomite, siderite, barytes and fluorspar, while ore minerals of economic importance may also be present.

Simple and Complex Veins.

The character of a mineral vein is controlled by the form of the original fracture, the manner of its opening, and the nature of the filling (see Fig. 5.5). A simple vein is found occupying a single well-defined fissure, with parallel or rectilinear walls, so forming a tabular or sheet-like body which can often be traced for a long distance. The internal structure may be massive if the mineral contents were distributed in a uniform manner throughout the vein in response to a single period of deposition as shown in Figure 5.5A. Alternatively, an internal banding may be developed parallel to the walls of the vein if successive layers differing in mineral content were deposited during its filling (see Fig. 5.5B). Although these layers are often rather irregular, they may be arranged symmetrically across the vein in such a way that the same sequence of layers is encountered outwards from its centre, as shown in Figure 5.5C. Such a banding is developed wherever a

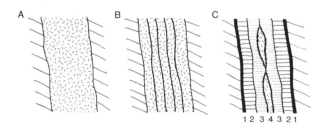

FIG. 5.5. Mineral veins and their structure.

fissure has opened repeatedly along its length. Since each layer is deposited on the walls of the fissure as it opened, the layers become younger towards the centre of the vein. These layers are commonly formed by prismatic minerals which grow with their long axes perpendicular to the walls of the fissure, giving rise to comb structure. It is the pyramidal terminations of these crystals, meeting across the centre of the vein to form a plane of weakness, which allow the fissure to open repeatedly along its length.

A complex vein is formed wherever the country-rocks or earlier vein-fillings have been fractured in several directions, giving rise to a zone of shattered rock which is traversed by a plexus of irregular fissures filled with mineral matter. Such a vein often has a brecciated structure, in that the angular fragments of shattered country-rock or vein-filling constitute a vein breccia. The mineral matter separating these fragments may be massive or banded, giving rise to cockade structure in the latter case.

Although the fissure formed by the opening of a fracture is usually filled completely with mineral matter, irregular cavities are found in some veins. These cavaties are known as vugs or druses. They are commonly lined with small crystals projecting from the walls, or they may be filled completely with mineral matter at a later stage.

Lodes and Vein Systems.

Mineral veins often occur in close association with one another, giving rise to a zone of closely spaced veins. This is known as a lode if it forms a workable ore deposit. The veins forming such a lode may be parallel to one another, so that the lode itself has a sheeted structure. Alternatively the lode may be formed by an anastomasing network of veins which branch and re-join in an irregular fashion along its length.

Mineral veins resemble joints in that they commonly occur parallel to one another, so forming a set of mineral veins. A system of mineral veins is developed if more than one set of mineral veins is present, differing in trend. Such systems of mineral veins are commonly shown on geological maps of a sufficiently large scale (say 1:50,000), particularly if the individual veins form workable ore deposits which have been explored in considerable detail. Commonly, two sets are found trending at right angles to one another, while subsidiary sets may be developed at oblique angles.

It is quite usual for the various sets of veins to differ in mineral content, even although there may be little evidence that they differ in age. Such evidence of relative age is obtained wherever mineral veins are seen to cut across and displace one another, since the vein so affected must be the earlier structure. However, this evidence only allows different sets of mineral veins to be dated in relation to one another if one set consistently affects the other set in the same way. This is not always the case.

Other Types of Mineral Veins.

Although some veins occupy fissures which opened at right angles to the original fracture, other veins occur in fissures that developed along faults. Such fissures can be formed during faulting if the movements are concentrated along an irregular fault plane. For example, a fault plane may show step-like changes in attitude as it passes through beds of differing lithology. Movement on such a fault plane allows fissures to open as shown in Figure 5.6A, which can then be occupied by mineral veins. Such cavities tend to be destroyed by continued movements so that veins of this type are usually only found along minor faults.

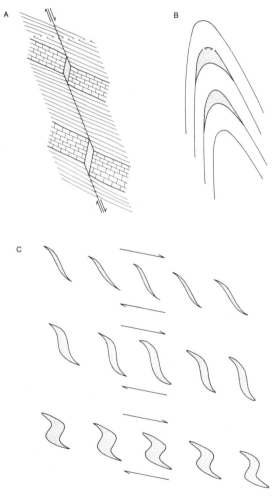

FIG. 5.6. A: Fissure vein formed along fault-plane by changes in dip on passing through different beds, B: Saddle-reefs, C: En echelon extension gashes.

Saddle-veins occupy lenticular fissures which tend to open along the hinge lines of folds in response to bedding plane slip. They are commonly found stacked one above the other along the axial plane of individual folds, as shown in Figure 5.6B. Ladder-veins are formed wherever an array of cross-cutting veins is developed along a narrow zone. Such an array could be developed if fracture is restricted to a competent layer such as a dyke. The veins may occupy extension-fractures, so that they form a single set of tension gashes. These may occur perpendicular to the layer, or they may be formed *en echelon* at an oblique angle to the layer. Alternatively, they could occupy shear-fractures, so forming conjugate sets at an acute angle to one another.

Tension gashes may also be developed *en echelon* along incipient shear zones. Such veins occupy extension fractures formed parallel to the direction of maximum compression, while the shear zone is developed at an acute angle to this direction. This allows the sense of movement on the shear zone to be determined, as shown in Figure 5.6C. Once formed, such a tension gash tends to rotate under the influence of the shearing movements, while it continues to propagate outwards across the shear zone with its original attitude, taking on a sigmoidal shape which is a direct reflection of the movements.

Faults and Fault-Planes

It has already been mentioned that a fault is a fracture along which there has been relative movement displacing the wallrocks on either side, as shown in Figure 5.7A. The essential feature is that the displacement must be accomplished, at least in part, by movement on a discrete surface so that the original continuity of the rocks is lost. This surface is known as the fault plane. The movements occur parallel to the fault plane. Although faulting

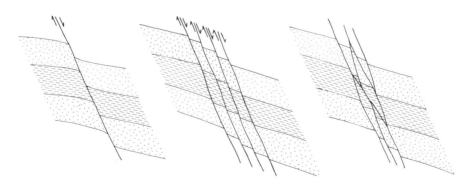

FIG. 5.7. Vertical cross-sections through faults differing in character from one another.
A: Single fault plane, B: Sheeted fault-zone; C: Braided fault-zone.

may result in a single fault plane, it is commonly the case that the movements take place on a series of closely spaced fault planes, giving rise to a zone of sheared, crushed and fractured rock affected by numerous small displacements. This is known as a fault zone. It may have a braided or sheeted structure (see Figs. 5.7 B and C). The wallrocks on either side of a fault are commonly fractured to a greater or lesser extent.

Individual faults can vary considerably in scale. Some are minor structures which only affect single beds of sedimentary rock, while others are major structures of regional importance, associated with large displacements of the earth's crust. Accordingly, some faults are only a few centimetres in length while others can be traced for hundreds or even thousands of kilometres. Likewise, the displacements on a fault can vary from less than a millimetre to much more than a hundred kilometres. It should be realised that geological maps only show those faults which can be seen to affect the outcrop pattern of the geological boundaries at a scale corresponding to the map itself. This means that only the larger faults are shown, with displacements of a few metres or more, depending on the scale of the map. The lines drawn as faults on such maps represent the intersection of the fault plane with the earth's surface. This intersection is termed the fault line or fault trace. Fault lines are commonly shown on geological maps as a particular type of geological boundary, so that they can easily be recognised.

Faults are often marked by topographic features which develop in response to differential erosion. Fault-line valleys and gullies may be eroded along fault lines, while fault-line scarps are formed where rocks differing in their resistance to erosion are brought into juxtaposition across fault lines. Offset ridges or escarpments are developed where layers resistant to erosion have been displaced across fault lines. All these features should be distinguished from the effects of recent faulting on the present-day topography, which vary according to the nature of the movements. Changes in elevation occur as a result of dip-slip movements on normal and reverse faults, while streams are offset by strike-slip movements on wrench faults.

Faults can be described and classified according to the attitude of the fault plane, the nature of the fault movements and the effects on the faulted rocks. Many of the terms applied to faults were originally introduced by miners. Some of these have become obsolete, while others have passed into the language of structural geology. The term "fault" had an origin in its use by miners to describe an interruption in the continuity of the strata being mined.

Attitude of the Fault Plane.

Since the fault plane is a surface, its attitude can be measured to determine the dip and strike at any point. Although it is natural for geologists to measure the dip as the angle from the horizontal by means of a clinometer, miners measured the complementary angle from the vertical, which was found using a plumb line (see Fig 5.8). This angle is known as the hade of the fault. The use of this term in structural geology is now obsolete. The attitude of the fault plane can be used to divide faults into high-angle faults (with dips of more than 45°) and low-angle faults (with dips of less than 45°). These terms are widely used.

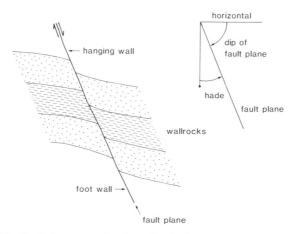

Fɪɢ. 5.8. Vertical cross-section through a fault showing the terms applied.

Faulting disrupts the original continuity of rock bodies to form a series of fault blocks in contact with one another. The rocks lying on either side of a fault plane define the walls of the fault blocks. If the fault is not vertical, the rocks lying above a fault plane define the hanging wall, so called because it was the wall where the miner hung his lamp, while the rocks lying below the fault plane define the foot wall, so called because it was the wall where the miner rested his foot, as shown in Figure 5.8. All these terms are used most appropriately to describe high-angle faults.

The attitude of a fault can be determined in three different ways. First, the fault plane may be exposed at the earth's surface, so that its dip and strike can be measured directly in the field. However, it is often the case that fault zones are less resistant to weathering and erosion than the adjacent rocks. This means that faults tend to be hidden under superficial deposits formed as a result of differential weathering and erosion, while their position may be marked by a topographic hollow, giving rise to a fault-line valley or gully. Secondly, the relief of the earth's surface may be sufficient to show the effect of the topography on the outcrop of the fault plane. It should then be possible to draw structure contours on the fault plane so that its attitude can be determined indirectly. Even if this cannot be done, it is often possible to distinguish between high-angle and low-angle faults because it is only the low-angle faults which are likely to be affected by the topography to any extent, unless the topographic relief is very marked. Such evidence is commonly shown by the outcrop pattern as illustrated by a geological map. Finally, subsurface data may be used to determine the attitude of a fault plane if its position at a certain depth can be compared with its outcrop at the earth's surface. This data may be obtained by drilling or mining. Once its position below the surface has been found, a simple application of the three-point problem is sufficient to determine the attitude of the fault plane. If there is sufficient data from below the surface, it may be possible to construct structure contours on the fault plane, thus allowing its outcrop at the earth's surface to be determined.

The attitude of a fault plane can also be considered in relation to the dip and strike of the rocks affected by the faulting, as shown in Figure 5.9. Thus, faults can be classified as: (1) dip faults trending parallel to the dip of the faulted rocks; (2) strike faults trending parallel to the strike of the faulted rocks; and (3) oblique faults trending at an oblique angle to the dip and strike of the faulted rocks.

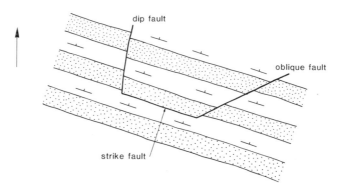

FIG. 5.9. Geological map showing relative attitude of dip, strike and oblique faults.

Additional terms are introduced to describe faults which affect folded rocks. Longitudinal faults strike parallel to the fold trend, and are typically found to affect the crests of anticlinal folds, as shown in Figure 5.10A. A common feature shown by such faults is that they tend to splay out at the plunging ends of the folds. They are often associated with transverse faults or cross faults, striking at a high angle to the fold trend, as shown in Figure 5.10B. Such faults may affect only the fold limbs, or they may cut across the fold hinges. They are often developed preferentially at the plunging ends of the folds. If the folding takes the form of domes and basins, longitudinal and transverse faults are replaced by tangential and radial faults, respectively. It can be noted that longitudinal and tangential faults are strike faults, while transverse and radial faults represent oblique and dip faults.

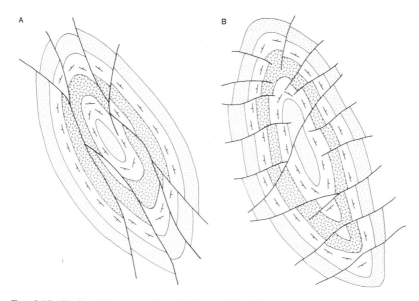

FIG. 5.10. Fault patterns associated with periclinal folds: A: Longitudinal faults, B: Cross and oblique faults.

Displacement Across Fault Planes

Faulting takes place whenever the rocks on either side of a fault plane are displaced relative to one another. By definition, the movements occur parallel to the fault plane, since they would otherwise result in a fissure. These movements can be described by the relative movement of two points which were originally adjacent to one another on either side of the fault plane, as shown in Figure 5.11. As the fault blocks move past one another, each point traces out a path on the opposite wall of the fault plane. This path is called the slip path. It is easier to visualise the slip path *AA'* generated on the foot wall by a point *A'* on the hanging wall, as shown in Figure 5.11. However, a slip path *BB'* is also generated on the hanging wall by corresponding point *B'* on the foot wall. Although the two paths are not the same, they are related to one another by symmetry, since the corresponding segments of each path are parallel to one another.

It should be clearly understood that there is generally no evidence to suggest which fault block has moved. Thus, the fault block on the left might have moved upwards to the right, generating the slip path on the hanging wall; or the fault block on the right might have moved downwards to the left, generating the slip path on the foot wall. This means that faulting can usually be described only in terms of relative movement. It is only the

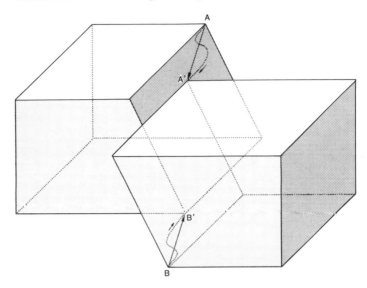

FIG. 5.11. Block diagram showing how the relative displacement is measured across a fault plane. The dotted lines AA′ and BB′ represent the actual displacements which took place on the fault plane while the straight lines AA′ and BB′ represent the net displacement which can be determined after the faulting has taken place. See text for further details.

evidence of the geological setting where faulting takes place which may allow an interpretation of the absolute movement to be attempted.

Overall Displacement.

The concept of a slip path can be used to describe the history of the movements which occur in sequence on a fault plane. However, it must be admitted that the evidence presented by an individual fault is rarely sufficient to define the slip path. Often, it is only possible to determine the overall displacement which has taken place. This is given by the net slip, found by drawing a straight line between the two points, on the walls of the fault plane, which were originally adjacent to one another. This straight line forms a chord joining the ends of the two slip paths, as shown in Figure 5.11. This line defines not only the direction of the displacement on the fault plane, but also its sense. Thus, one end of the line represents the initial position *A* of a point, while the other end represents its final position *A′*. The sense of overall displacement is then given by the line *AA′*, as shown by the arrow joining *A* to *A′* in Figure 5.11.

Translational and Rotational Movements.

This discussion has dealt with the slip that has taken place at a particular point on a fault plane. However, it is possible to distinguish two types of fault, according to whether or not the net slip on a fault plane varies from point to point. Translational faults occur if the net slip is the same for every point on the fault plane, so that parallel lines on either side of the fault remain parallel to one another after the faulting has taken place, as shown in Figure 5.12A. Such faults can be compared with rotational faults which occur whenever the net slip varies from point to point on the fault plane, as shown in Figure 5.12B. If the rocks forming the walls of the fault plane act as rigid blocks, the movements can only be the result of rotation about an axis which lies at right angles to the fault plane. The net slip increases away from the point where this axis intersects the fault plane, while the slip paths form circular arcs centred on this point. The faulting affects parallel lines lying on either side of the fault plane, so that they are no longer parallel to one another after the movements have taken place. The only exception to this statement concerns lines parallel to the axis of rotation, which remain parallel to one another.

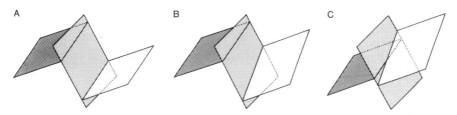

Fig. 5.12. Block diagram showing different types of fault. A: Translation fault. B: Rotational fault. C: Scissor fault.

This distinction between translational and rotational faults is rather arbitrary. It is likely that the displacement on most faults varies to a certain extent, so that there is always a component of rotational movement to be considered. Moreover, all faults must end as they are traced along their strike, with the sole exception of ring faults. A component of rotational movement would also be developed wherever a fault dies out, rather than terminating against another fault which can take up the displacement. Since the end of such a fault acts as a hinge, it is known as a hinge fault. Another type of rotational fault is developed wherever the sense of displacement is reversed across a point of zero slip, while the amount of displacement increases in either direction away from this point. Such a fault is known as a scissor fault, as shown in Figure 5.12C.

Distortion of the Country-rocks.

It has been assumed so far that faulting affects rigid blocks of rock. However, the recognition of hinge faults casts doubt on this assumption, since the displacements on such a fault must result from the bending or warping of the faulted rocks about a hinge line. In fact, it is generally the case that rocks affected by faulting are distorted to a varying extent. This can occur in two ways.

Firstly, the rocks against a fault plane may be distorted by the frictional drag exerted by the fault movements. For example, flexures are often developed in sedimentary rocks where fault drag results in the deflection of the bedding planes against the fault plane, as shown in Figure 5.13A. In fact, it is not certain that these flexures are the result of fault drag, since sliding friction on a fault plane is usually less than the internal friction which controls the resistance of rocks to fracture and flow. It is equally possible that they developed at an early stage of the movements, prior to the formation of the fault itself. Such flexures provide some evidence about the sense of displacement on a fault plane. However, it should be recognised that the slip direction is unlikely to lie at right angles to the hinge of the flexure, as might appear to be the case. Moreover, although the sense of displacement is shown by the bending of the beds as they approach the fault plane, it is sometimes found that flat-lying strata are affected by "reverse drag". This means that the beds bulge upwards on the upthrown side, while the opposite effect occurs on the downthrown side, as shown in Figure 5.13B. The folding and disruption associated with faults means that observations concerning the attitude of bedding must be treated with caution whenever they are obtained near fault zones.

Secondly, the rocks lying between fault planes are commonly affected by differential movements, giving rise to wraps, sags and flexures. It is often the form of these undulations which control the amount of displacement on the fault planes. For example, the displacements on a fault which only extends for a limited distance before dying out must result from the distortion of the rocks lying on either side of the fault plane.

Slickensides.

Fault planes commonly occur in the form of slickensided surfaces which have been polished and striated by the fault movements. The term "slicken-

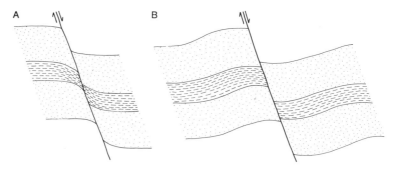

FIG. 5.13. Distortion of wall rocks as a result of: A: Normal fault-drag. B: Reverse fault-drag.

side" was originally applied to the polished nature of the surface, formed by a thin film of comminuted and mineralised material which lines the fault plane. However, it now refers more to the striations developed on such a polished surface than the surface itself. These striations form parallel to the direction of displacement on a fault plane. Since the striations formed during the earlier stages of faulting are likely to be destroyed by the later movements, they provide only a partial record of the displacements on the fault. Generally, they give the direction of the latest movements on the fault plane. There is no certainty that this corresponds to the direction of the earlier movements. Occasionally, several sets of slickensides are developed at an acute angle to one another, indicating that slight changes in the direction of displacement took place towards the end of the fault movements.

Similar features are also formed by the fibrous growth of minerals such as quartz between the walls of the fault plane. This growth implies that slight gaps develop during the faulting. This may occur because the fault plane is slightly irregular, so that movement tends to open up spaces as shown in Figure 5.14. If the mineral fibres can be traced across the whole fault zone without any break, it is possible to reconstruct the entire history of the fault movements. The fibres may be curved, indicating that the direction of displacement has changed during the course of the fault movements.

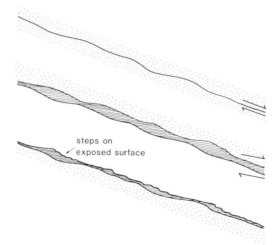

Fig. 5.14. Fibrous growth of minerals along a fault plane, associated with the development of slickensides.

Exposed surfaces in such a fault zone are commonly stepped at right angles to the direction of displacement given by the slickensides. These steps are formed by the rock breaking along fractures which cut across the mineral fibres as shown in Figure 5.14. The steps on one fault-block face in the same direction as the other fault-block has moved. Even if these steps are too small to be noticeable, the fault plane may still feel smooth if it is stroked by the fingers in the direction of movement of the other block, whereas it feels rough in the opposite direction. Although this way of determining the sense of displacement on a fault plane has been applied in the past to any slickensided surface, experimental evidence has suggested that it may not always be reliable. The method is best applied to fault surfaces where the fibrous nature of the slickensides is clear.

Fault planes are occasionally grooved or fluted, giving rise to large corrugations with a wavelength of one or two metres, and a depth of several centimetres. Such corrugations are formed parallel to the direction of displacement.

Feather Joints and Splay Faults.

Faulting may also cause fractures which affect the wallrocks on either side of the fault plane. Although these fractures differ in kind, they commonly occur as single sets of parallel structures, which lie at an acute angle to the main fault. Feather joints are formed by a single set of extension joints arranged *en echelon* along the fault plane. The acute angle of less than 45° between these joints and the fault itself gives the sense of displacement, as shown in Figure 5.15A. Shear fractures may also be developed, generally as a conjugate set of minor faults which lie at an acute angle of more than 45° to the main fault (see Fig. 5.15B). These faults have a sense of displacement which is opposed to the movements on the main fault. Finally, minor faults may also occur as splay faults diverging from the main

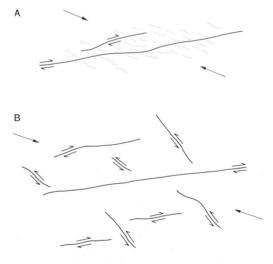

FIG. 5.15. Extension and shear fractures associated with a major fault-plane: A: Feather joints (dotted lines). B: conjugate fault systems.

fault at an angle of less than 45°. These faults have the same sense of displacement as the main fault. Since they are commonly developed near the ends of a major fault, they may allow the displacement on such a fault to die out.

Classification of Faults

The concepts introduced in the previous sections can now be used to classify fault movements. This relates the slip direction and the sense of overall displacement to the attitude of the fault plane. It serves as the basic classification for describing different types of fault.

Strike-slip Faults.

If the slip direction is parallel to the strike of the fault plane, the fault is termed a strike-slip fault. It is usually a high-angle fault with a vertical fault plane, as shown in Figure 5.16. Such faults are commonly known as wrench faults, transcurrent faults, tear faults or transform faults. These names have genetic connotations which will be discussed later in the next chapter.

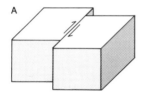 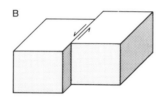

FIG. 5.16. Block diagram illustrating the two types of strike-slip faults: A: Right-lateral (dextral) strike-slip fault. B: Left-lateral (sinistral) strike-slip fault.

The sense of overall displacement on a strike-slip fault is determined by an observer looking across the fault line. If the rocks across the fault appear to have moved to the left, the fault is known as a left-lateral (or sinistral) strike-slip fault. Alternatively, if the rocks across the fault appear to have moved to the right, the fault is known as a right-lateral (or dextral) strike-slip fault. It should be clearly understood that the sense of overall displacement does not depend on which side of the fault plane the observer stands. The rocks on the other side always appear to have moved left or right, as the case may be.

Dip-slip Faults.

If the slip-direction is parallel to the dip of the fault, as shown in Figure 5.17, the fault is termed a dip-slip fault. The sense of overall displacement on a dip-slip fault is determined according to the relative movement of the

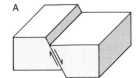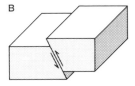

FIG. 5.17. Block diagram illustrating the two types of dip-slip faults: A. High-angle normal dip-slip fault. B. High-angle reverse dip-slip fault.

hanging wall to the foot wall. if the hanging wall appears to have moved downwards, the fault is known as a normal dip-slip fault. Alternatively, if the hanging wall appears to have moved upwards, the fault is known as a reverse dip-slip fault. Since the use of the terms "normal" and "reverse" implies that the fault is a dip-slip fault, it is common practice to refer to such faults simply as normal or reverse faults.

These terms were originally introduced by coal-miners. Normal faults tend to displace coal-seams downwards in the same direction as the dip of the fault plane. Accordingly, if such a fault was encountered in the working of a coal seam on the upthrown side, it was normal practice to continue the heading for a short distance, and then sink a shaft to regain the coal-seam at a lower level beyond the fault. If the fault had the opposite sense of overall displacement, this procedure would not work. The reverse of the normal practice had then to be followed to regain the coal-seam.

This classification of dip-slip faults depends on the attitude of the fault plane, as shown in Figure 5.18. For example, if the fault plane passes through either the vertical or the horizontal, a normal fault changes into a reverse fault, and vice versa. This may occur along the strike, so that the fault plane appears to be twisted into a horizontal helix, or it may occur down the dip, so that the fault plane has a curved form.

Normal faults are commonly high-angle faults with fault planes dipping at angles of 60° or 70°, as shown in Figure 5.18 A. Such faults are known simply as normal faults, since the high-angle nature of the fault plane is taken for granted. However, if a normal fault has a fault plane which dips at less than 45°, as shown in Figure 5.18D, it is usually distinguished by name as a low-angle normal fault.

Reverse faults occur as high-angle faults or low-angle faults, as shown in Figures 5.18B and 5.18C. It is common for the fault plane to curve in the direction of the dip so that a high-angle reverse fault changes into a low-angle reverse fault or vice versa (see Figure 5.19A). These qualifying terms should always be applied. High-angle reverse faults are known as upthrusts, while low-angle reverse faults are commonly called thrust faults (thrusts) or overthrust faults (overthrusts). The use of the latter term at least implies that the dip of the fault-plane is very low. Indeed, it is quite common for overthrusts to be folded so that the fault plane passes through the horizontal in the direction of dip, as shown in Figure 5.19C. Strictly speaking, this

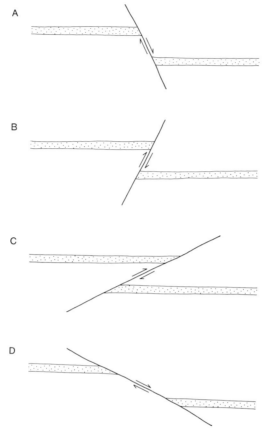

Fɪɢ. 5.18. Vertical cross-sections showing how the dip of the fault-plane affects the nature of dip-slip faulting. Note that the downthrown fault-block always occurs to the right so that the sense of displacement as given by the fault arrows always remains the same: A: High-angle normal fault. B: High-angle reverse fault or upthrust. C: Low-angle reverse fault or overthrust. D: Low-angle normal fault.

would mean that a low-angle reverse fault (or overthrust) has changed into a low-angle normal fault. However, it is customary to apply the term "over-thrust" to such faults as a whole.

Even so, several objections can be raised against the use of this term. Firstly, although it implies an origin by the thrusting of the upper layers over the lower layers, it is equally possible that the lower layers were thrust under the upper layers. It would then be an underthrust. However, there is rarely sufficient evidence to decide whether overthrusting or underthrusting took place. Usually, it is only the nature of the relative movements which can be determined. Secondly, an origin by overthrusting (or underthrusting) becomes less certain as the fault plane approaches the horizontal. It is equally possible that "overthrusting" occurs in response to gravity sliding

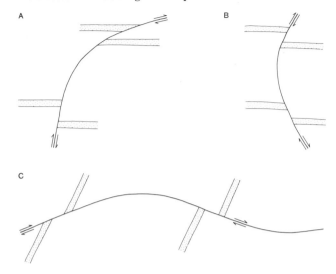

FIG. 5.19. Vertical cross-sections showing relationships between different types of dip-slip faults: A: High-angle reverse fault passing upwards into low-angle reverse fault. B: High-angle normal fault passing upwards into high-angle reverse fault. C: Low-angle reverse fault or overthrust passing across strike into low-angle normal fault.

or lateral spreading. Thus, the term not only implies a sense of absolute movement which is difficult to substantiate but also suggests a mode of origin which is only one of several possible hypotheses. However, it is a term which is deeply entrenched and commonly used in the literature of structural geology. It should be emphasised that its use only implies over-thrusting in the sense that one layer is brought to rest on top of another.

Oblique-slip Faults.

This classification of faults only recognises dip-slip or strike-slip movements on the fault plane. However, oblique-slip movements can also occur at an oblique angle to the dip and strike of the fault plane. Such faults are known as oblique-slip faults. Since the overall displacement on the fault plane has dip-slip and strike-slip components, inspection of Figure 5.20 shows that oblique-slip faults can be classified as follows:

    (A)   Left-lateral normal oblique-slip faults.
    (B)   Right-lateral normal oblique-slip faults.
    (C)   Left-lateral reverse oblique-slip faults.
    (D)   Right-lateral reverse oblique-slip faults.

The relative importance of the dip-slip and strike-slip components on an oblique-slip fault can be related to the angle that the slip direction makes with the horizontal, as measured in the fault plane. This angle gives the pitch of the slip direction in the fault plane. Rake is an obsolete term which was also applied to this angle.

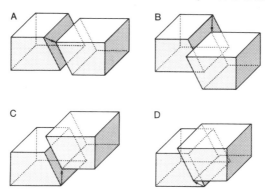

FIG. 5.20. Block diagrams showing the four types of oblique-slip faults. See text for further details.

Assumption of Straight-line Slip-paths.

It should be emphasised that this general classification of faults assumes that the slip-path is a straight line corresponding to the overall displacement. This assumption can be justified in general by considering the consequences of a curved slip-path. This would mean that the slip direction changed during the course of the faulting so that the fault movements would be oblique-slip. Since dip-slip and strike-slip faults would only then be developed under special circumstances, oblique-slip faults should be more common than dip-slip or strike-slip faults. However, the opposite appears to be the case, suggesting that slip-paths are usually straight lines corresponding to the overall displacements on fault planes. Moreover, since these straight-line paths result in either dip-slip or strike-slip faults, the slip directions appear to be preferentially developed parallel to either the dip or the strike of the fault planes. This assumption is less likely to be correct if the fault plane has been affected by rejuvenation, so that it forms a plane of weakness on which faulting can be renewed.

Effect of Faulting on Sedimentary Rocks

Fault lines are commonly shown as thick, dashed or coloured lines on a geological map in order to distinguish them from the other type of geological boundary. However, they also introduce discontinuities into the outcrop pattern of the geological formations, which would allow them to be distinguished as a particular type of geological boundary. Such discontinuities show how the faulting has affected the outcrop of the geological formations by the way that the geological boundaries between these formations have been shifted across the fault lines. Although these boundaries may represent any sort of geological boundary, they are typically formed by the contacts between sedimentary formations.

Use of Structure Contours.

It has already been shown how structure contours can be drawn on a stratigraphic contact to illustrate its form in three dimensions. The intersection of these contours with the topographic contours defines the outcrop of the geological boundary corresponding to this contact at the earth's surface. It is also possible to draw structure contours on a fault-plane to show its form in three dimensions. The intersection of these contours with the topographic contours then defines the outcrop of the fault line at the earth's surface.

Now consider the effect of faulting on a stratigraphic contact. This will result in the contact being shifted across the fault plane. If structure contours are drawn on this contact, as shown in Figure 5.21, they show a discontinuity which represents the position of the fault plane. This disconti-

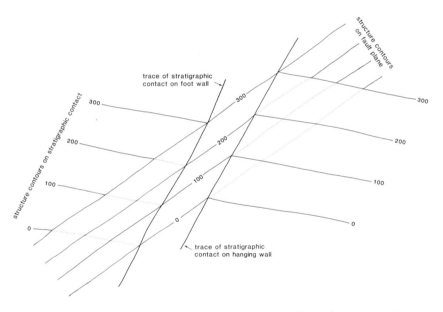

FIG. 5.21. Diagram showing how the traces made by a stratigraphic contact against a fault plane can be determined using structural contours. The lines trending approximately east-west represent structure contours drawn on the stratigraphic contact while the lines trending approximately north-east-south-west represent structure contours drawn on the fault plane itself. See text for further details.

nuity is marked by each contour ending where the stratigraphic contact abuts against the fault plane. The traces made by the stratigraphic contact on the walls of the fault plane can then be found by drawing structure contours on the fault plane. The structure contours drawn at equivalent heights on both the stratigraphic contact and the fault plane will then intersect one another in a series of points which lie along two lines. Each line corresponds to the trace made by the stratigraphic contact against the fault plane. Together these lines define the subsurface "outcrop" or subcrop of the stratigraphic contact on the walls of the fault plane. As such, they define the effect of faulting on the stratigraphic contact.

It should be clearly understood that the structure contours drawn at the same height on the stratigraphic contact on either side of the fault plane have no intrinsic relationship to one another. In other words, they certainly do not need to represent lines inscribed on the stratigraphic contact which were originally continuous with one another. This would only be the case if the net slip on the fault plane was the result of strike-slip movements. Corresponding points on either side of the fault plane would then remain at the same height while the movements were taking place. However, inspection of Figure 5.22 shows that exactly the same geometry can be developed

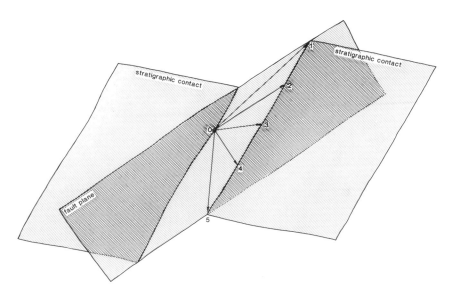

FIG. 5.22 Diagram showing how the different types of fault displacement appear to produce exactly the same effect on a stratigraphic contact. Heavy arrows denote slip paths as follows: 1. Left-lateral reverse oblique-slip fault; 2. left-lateral strike-slip fault; 3. left-lateral normal oblique-slip fault; 4. normal dip-slip fault; 5. right-lateral normal oblique-slip fault.

by overall displacements which range through 180° of arc. Such displacements show differing components of dip-slip and strike-slip movements, as defined by the varying pitch of the overall slip-direction within the fault plane. In particular, it can be seen from Figure 5.22 that an identical geometry may be developed by dip-slip movements as by strike-slip movements. Accordingly, a planar surface such as a stratigraphic contact provides no evidence whereby the net slip across a fault plane can be determined; it is simply not possible to identify which points on the contact were originally adjacent to one another.

Separation and its Definition.

The effect of faulting on a stratigraphic contact is generally described in terms of the separation shown by the two segments of the faulted surface on either side of the fault plane. Separation is simply a measure of the distance between those segments in a particular direction. This direction must be specified. For example, consider the traces made by the faulted surface on the walls of the fault plane, as shown in Figure 5.23. The strike separation is then given by the distance between these traces as measured parallel to the strike of the fault plane. The dip separation is given by the corresponding distance parallel to the dip of the fault plane.

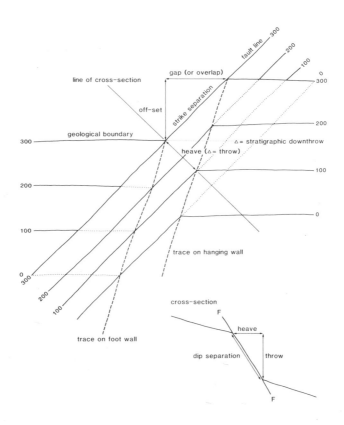

FIG. 5.23. Geological map and a vertical cross-section illustrating the various terms applied to the separation of a stratigraphic horizon across a fault plane. Δ = difference in height as measured using the structure contours.

The strike separation affecting a stratigraphic contact can be determined directly from the outcrop pattern shown by a geological map wherever the topography is horizontal. If the fault plane has been contoured, the strike separation is given by the distance between the traces made by the faulted surface on the fault plane, as measured parallel to the structure contours drawn on the fault plane (see Fig. 5.23). Its character can be described as left-lateral (sinistral) or right-lateral (dextral) according to the direction where the contact reappears on the far side of the fault line. The strike separation shown by a geological map can be divided into two components, as shown in Figure 5.23. The component of strike separation measured parallel to the dip of the faulted surface is known as the offset of the stratigraphic contact. The component of strike separation measured normal to this direction is known as the overlap or gap of the stratigraphic contact depending on the nature of the outcrop pattern.

The dip separation affecting a stratigraphic contact can be seen in a vertical cross-section drawn at right angles to the strike of the fault plane. Its character can be described as normal or reverse depending on whether the trace made by the faulted surface on the foot wall is higher or lower than the corresponding trace on the hanging wall. Whether or not this would result in the vertical repetition of the faulted surface depends on the circumstances, as shown in Figure 5.24. If it does, the same horizon would be encountered twice in a vertical borehole which was drilled between the traces made by the faulted surface against the fault plane. Otherwise, this horizon would not be encountered by such a borehole, since it is not present at depth between these traces.

The dip separation can be divided into two components. The vertical component may be considered to define the throw of the fault, as it affects a particular horizon. It is given by the difference in height between the traces made by the faulted surface on the fault plane, as measured in a direction parallel to the dip of the fault plane. The throw can be determined wherever structure contours have been drawn on the faulted surface, as shown in Figure 5.23. The horizontal component of the dip separation may be considered to define the heave of the fault, as it affects a particular horizon. It is given by the horizontal distance between the traces made by the faulted surface on the fault plane, as measured in a direction parallel to the dip of the fault plane.

Although there has been some confusion in the literature about the meaning that should be attached to the terms "throw" and "heave", they are best taken to describe the effect of the faulting on a particular horizon. They are therefore defined in relation to the separation shown by this horizon across the fault plane. So defined, these terms do not refer to the displacement that has occurred on the fault plane. This can be made clear by using "apparent throw" and "apparent heave" wherever this is thought to be necessary.

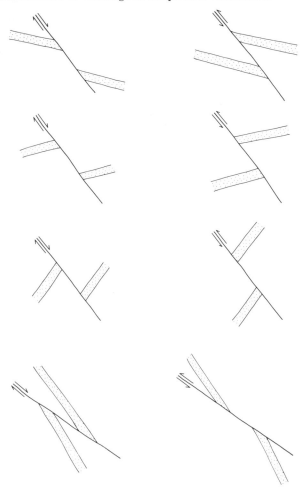

FIG. 5.24. Vertical cross-sections showing differences in dip separation across a fault plane, according to the dip of the faulted horizon relative to the fault plane.

The down-dip method of viewing the outcrop pattern can be used to determine the nature of the dip separation shown by a fault. To do this, the geological map is viewed at an oblique angle along the fault line, corresponding to the dip of the faulted surface in the same direction. Such a foreshortened view of the outcrop pattern can then be regarded as equivalent to a cross-section. It shows that the faulted horizon is closer to the observer on the downthrow side of the fault line, so allowing the downthrow on the fault plane to be determined. If the dip of the fault plane is known, the dip separation can also be classified as normal or reversed, according to whether the downthrow side forms the footwall or the hanging wall of the fault plane.

Stratigraphic Separation.

If faulting affects sedimentary rocks, the separation across a fault plane can also be considered in relation to the stratigraphic horizons which are brought into contact with one another. These horizons would be separated from one another by a certain thickness of intervening strata. The true thickness of these beds gives the stratigraphic separation across the fault-plane. It is measured normal to the bedding. The stratigraphic separation can therefore be determined wherever the horizons brought into contact across the fault plane can be identified with sufficient accuracy that the thickness of the intervening beds can be found from a detailed knowledge of the stratigraphic column. This method can be applied even if the beds are folded and the topography is not horizontal.

The stratigraphic separation can be resolved to give what can be termed the stratigraphic downthrow, which measures the separation shown by the faulted surface in a vertical direction. It is determined from the difference in height of the faulted surface across the fault plane, as measured parallel to the strike of the faulted surface.

It should be emphasised that the stratigraphic downthrow does not provide the same measure of separation as the apparent throw across the fault plane. The stratigraphic downthrow is measured parallel to the strike of the faulted surface, while the apparent throw is measured parallel to the dip of the fault plane. These two directions are not equivalent to one another. However, there are three cases where these components of separation are equal to one another: (1) if the faulted surface strikes at right angles to the fault plane, to give a dip fault; (2) if the fault plane is vertical; or (3) if the faulted surface is horizontal. The stratigraphic downthrow and the apparent throw would differ most markedly from one another if the faulted surface has the same strike as the fault plane, to give a strike fault.

Since it is often not possible to determine the attitude of a fault plane, it is usually the stratigraphic downthrow which is measured, using the structure contours drawn on the faulted horizon. It is the difference in height between these contours on either side of the fault line, which gives the stratigraphic downthrow. However, it should be realised that the stratigraphic downthrow would not be very different from the apparent throw in the case of a high-angle fault affecting flat-lying beds. For example, if a strike fault dipped at 60° in the same direction as beds dipping at 30°, the stratigraphic downthrow would be two-thirds of the apparent throw. This difference would be reduced if the fault approached the vertical, if the beds approached the horizontal, or if the beds dipped in the opposite direction to the fault plane.

Determination of the Net Slip.

It has already been emphasised that the actual displacement or slip on a fault plane cannot be determined from the separation shown by a single plane. However, this becomes possible if two discrete surfaces are affected by the faulting, so that they intersect one another in a line which has a certain position in space. The traces made by these surfaces against the fault plane are shown in Figure 5.23, which is a diagram of the fault plane as viewed at right angles to itself. The solid lines represent the traces of the faulted surfaces on the foot wall, while the dashed lines give the corresponding traces on the hanging wall. Assume that the traces AB and CD on the foot wall intersect one another in the point X, while the corresponding traces $A'B'$ and $C'D'$ on the hanging wall intersect one another in the point X'. The two surfaces intersect one another in a line which is displaced across the fault plane so that it impinges on either wall at the points X and X'. Since X and X' were originally adjacent to one another on either side of the fault plane, the net slip-path of the hanging wall relative to the foot wall is given by the line XX'. The displacement shown in Figure 5.25 corresponds to a normal left-lateral oblique-slip fault. Such a diagram can be constructed graphically using the stereographic projection if the dip, strike and relative position of the various surfaces are known, although the method is beyond the scope of this book.

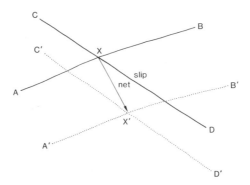

Fɪɢ. 5.25. Diagram showing the net slip can be determined on a fault plane from the traces made by two discrete surfaces. see text for further details.

It should be emphasised that the method involves finding the net displacement of a line across a fault plane, even although the line may be defined by the intersection of two surfaces. It is the impingement of this line on the walls of the fault plane which serves to identify points which were originally adjacent to one another before the faulting took place. A variety of lines defined by structural or stratigraphic relationships can be used to determine the net slip on a fault plane. These include the following:

1.   Lines formed by the intersection of two planes (including the traces made by one plane against another):
- (a)   Intersection of sheet intrusions or mineral veins with one another.
- (b)   Intersection of a sheet intrusion of mineral vein with a stratigraphic or ingenous contact.
- (c)   Traces made by an igneous contact cutting across stratigraphic contacts, igneous intrusions, mineral veins, faults and folds.

2.   Lines formed by structural and stratigraphic feature of a linear nature:
- (a)   Hinges, crests and troughs of folds.
- (b)   Sedimentary channels, shoe-string sands, off-shore bars and wash-outs.
- (c)   Lateral changes in the facies of sedimentary rocks.
- (d)   Stratigraphic feather-edges formed by the overlap of sedimentary formations.
- (e)   Isopachytes as lines of equal sedimentary thickness.
- (f)   Ore-shoots, volcanic plugs and necks, and plutons

Although many of these features have yet to be described, they are listed here for the sake of completeness. It is found in practice, however, that the character of a fault can be determined most easily if folded rocks are affected.

Effect of Faulting on Folded Rocks

The net displacement on a fault can be determined most easily from its effect on the outcrop pattern of folded rocks because the separation across the fault plane varies according to the attitude of the folded surface. Figure 5.26 shows an antiform inclined to the south-west, so that its south-western limb dips more steeply than its north-eastern limb, while the fold plunges at a gentle angle to the north-west. The axial plane dips steeply towards the north-east. It is cut by a fault with a moderately steep dip towards the east-south-east, trending towards the north-north-east.

It can be seen that the effects of left-lateral strike-slip movements on this fault would develop an outcrop pattern as shown in Figure 5.26A. The south-western limb shows normal dip-separation combined with left-lateral strike-separation, while the north-eastern limb is affected by reverse dip-separation combined with left-lateral strike-separation. Right-lateral strike-slip movements have the opposite effect on the outcrop pattern, as shown in Figure 5.26C. The south-western limb shows reverse dip-separation combined with right-lateral strike-separation, while the north-eastern limb is

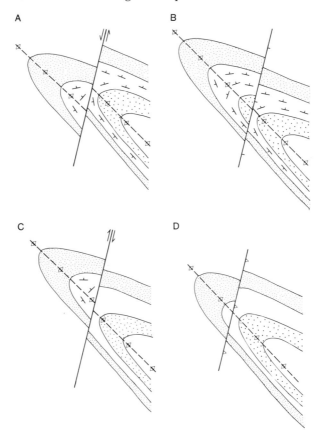

FIG. 5.26. Effect of faulting on folded rocks: A: left-lateral strike-slip fault. B: Normal dip-slip fault. C: Right-lateral strike-slip fault. D: Reverse dip-slip fault. The fault plane is assumed to dip in an easterly direction.

affected by normal dip-separation combined with right-lateral strike-separation. In other words, the strike separation remains the same, while the nature of the dip separation changes along the fault line, according to the dip of the folded rocks. The outcrop of a formation lying in the core of the fold does not change in width as a result of strike-slip faulting, while the position of the axial trace is affected if the fold is relatively upright. The apparent downthrow across the fault plane changes systematically along the line of the fault, so that the fold limbs are downthrown in opposite directions. This means that the fault appears to be rotational in character. However, it should be clearly recognised that such a change in downthrow is not the result of scissor faulting. Such an explanation would require that the corresponding fold-limbs on either side of the fault line differ in attitude, for which there is no evidence in the present case.

The effect of dip-slip movements on the outcrop pattern can now be considered. Figure 5.26B shows the outcrop pattern that is developed as a result of normal dip-slip movements on the fault plane, downthrowing the fault block to the east. The south-western limb shows normal dip-separation combined with left-lateral strike-separation, while the north-eastern limb is affected by normal dip-separation combined with right-lateral strike-separation. Reverse dip-slip movements resulting in the downthrow of the fault block to the west would have the opposite effect on the outcrop pattern, as shown in Figure 5.26D. The south-western limb shows reverse dip-separation combined with right-lateral strike separation while the north-western limb is affected by reverse dip-separation combined with left-lateral strike separation. In other words, the dip-separation remains the same, while the nature of the strike-separation changes along the line of the fault, according to the dip of the folded rocks. The outcrop of a formation lying in the core of the fold changes in width across the fault line as a result of the dip-slip movements, while the position of the axial trace is not greatly affected if the fold is relatively upright.

It should be recognised that, while the nature of the strike separation can be found directly from an examination of the outcrop pattern, the dip separation can only be determined as normal or reverse if the direction of dip of the fault plane is known. The evidence presented by a geological map often does not allow this to be determined, so that the true nature of the dip separation cannot be found. However, it is still possible to determine whether the faulting is a result of dip-slip or strike-slip movements according to whether or not the strike separation changes in sense on tracing the fault line from one fold limb to another. Only if the direction of dip shown by the fault plane can be determined is it then possible to identify a dip-slip fault as normal or reverse. By way of contrast, the classification of a strike-slip fault as left-lateral or right-lateral does not depend on the inclination of its fault plane, so allowing such a fault to be identified without any uncertainty.

The effect of faulting on the outcrop pattern of folded rocks can only be used to distinguish a dip-slip fault from a strike-slip fault if the fault cuts across a fold hinge. However, the offset across a dip-slip fault decreases to zero as the dip of the bedding increases to the vertical, while the offset across a strike-slip fault always remains the same, provided that the bedding is not horizontal. This means that recognisable faults in steeply dipping rocks are more likely to be strike-slip faults, while faults in flat-lying rocks are more likely to be dip-slip faults. It should be clearly recognised that there are many exceptions to this statement.

Diagnostic Features of Fault Patterns.

Faults are classified according to the net displacement which has taken place on the fault plane, as judged in relation to its attitude. It is relatively easy to distinguish low-angle faults from high-angle faults according to the way the topography affects the outcrop pattern. However, it is more difficult to determine the net displacement on a high-angle fault even although it is this parameter which serves to distinguish strike-slip faults from normal and high-angle reverse faults with dip-slip movements. It is also difficult to determine the direction of dip shown by a high-angle fault even although it is this parameter which serves to distinguish normal faults from high-angle reverse faults. This means that the character of a particular fault can often only be determined by considering the nature of the fault pattern and its tectonic setting.

In general, faults tend to occur as sets of parallel fractures while fault sets often form fault systems consisting of conjugate sets of complementary shear fractures arranged at an acute angle to the direction of maximum compression. It is the varying arrangement of such fault sets which gives rise to the fault patterns now to be described. All this evidence must be borne in mind when attempting to interpret the outcrop pattern shown by a geological map.

Normal Faults

Ideally, normal faults from conjugate sets of high-angle fractures on which dip-slip movements with normal displacements have taken place. This means that the hanging wall has moved downwards relative to the foot wall. These fractures intersect one another in a horizontal direction, so that they tend to strike parallel to one another while they dip in oppossite directions. The downthrow on these faults occurs in the direction of dip.

Such a fault system develops in response to a vertical direction of maximum compression, assuming that this is associated with a horizontal direction of minimum compression, as shown in Figure 5.27. Since the ver-

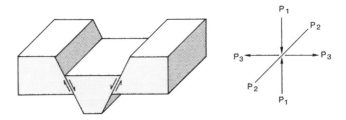

Fig. 5.27. Conjugate nature of normal faulting.

tical compression is caused by the weight of the overlying rocks, such faults have been termed gravity faults. However, as the vertical stress is effectively constant at a given depth, it is a reduction in the horizontal stress which should be considered as the cause of normal faulting. This develops in response to a horizontal tension, superimposed on the usual state of compressive stress which acts at depth. Although tensile stresses are unlikely to occur at any depth in the earth's crust, such a stress system is generally termed a horizontal tension whenever the horizontal stress is less than the vertical. It should be noted that any system of normal faults result in the extension of the earth's crust in a horizontal direction.

While normal faults can occur in any tectonic setting, they are typically found in areas of sedimentary rocks which are otherwise not affected by folding or faulting to an appreciable extent. This means that it is normal faults which are usually developed in areas of flat-lying or gently-folded sediments. Since other fault systems rarely occur in such a setting, it may be considered characteristic of normal faults.

Characteristic Features.

The fault patterns developed by normal faults where they affect flat-lying or gently-folded sequences of sedimentary rocks show several characteristic features, as shown in Figure 5.28. Although these faults often trend in the same direction, the fault lines are rarely straight. Instead, individual faults tend to shown slight changes in direction, giving rise to fault lines which are somewhat irregular in trend. This tendency may be accentuated so that the fault lines show gradual or abrupt changes in direction along their lengths. It is possible that such variations in the trend of normal faults reflects the structure of the underlying rocks. The ends of normal faults are often curved where they die out. It is commonly found that the rocks lying between major faults are cut by a series of minor faults, which often show

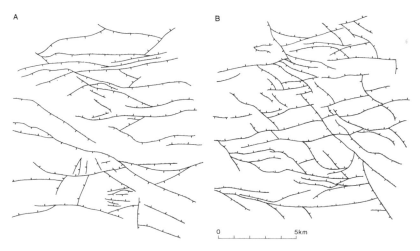

FIG. 5.28. Typical patterns developed by normal faults in flat-lying or gently folded rocks: A: Single trend dominant. B: Two distinct trends, possibly of different age.

wide variations in trend. These minor faults usually terminate against the major faults. If the major and minor faults are equally well developed, a fault pattern consisting of a complex mosaic of normal faults may result. The termination of one fault against another, which is a common feature of such patterns, is easily understood from geometrical considerations. The displacements must be parallel to the direction in which these faults planes intersect one another. This means that the dominant movements will be dip-slip in character whenever high-angle fault planes are involved.

Such complex patterns may be formed in response to a single episode of normal faulting. However, it is equally possible that such movements have occurred more than once during the course of geological history. The only evidence clearly supporting this hypothesis is shown by fault patterns formed by intersecting sets of normal faults which have distinctive trends at an oblique angle to one another. It is argued that the earlier faults form planes of weakness in the rock, along which the later movements are deflected. Accordingly, it is the earlier faults which are more continuous and regular in their development, while it is the later faults which appear to be affected by the earlier faults, curving as they approach and terminate against these pre-existent planes of weakness. These relationships are the opposite of those which might be expected at first sight. However, it must be admitted that the evidence of this sort must be treated with caution since it is often contradictory.

Step Faults and Block Faults.

So far, we have only considered the pattern made by normal faults at the earth's surface. However, the effect of the faulting also depends on whether or not the fault planes all dip in the same direction. Normal faults trending in the same direction may therefore form step faults or block faults, according to the nature of the downthrow, as shown in Figure 5.29.

Step faults are formed wherever a single set of normal faults is developed parallel to one another, so that the downthrows are all in the same direction. The sedimentary rocks lying between these faults are commonly tilted so that they dip in the opposite direction, giving rise to tilted fault blocks. The tilting may develop as a result of the fault plane becoming less inclined at depth, since this would allow rotational movements to affect the intervening fault blocks.

Block faults are formed by conjugate sets of normal faults, dipping in opposite directions to one another. The downthrows are such that the rocks lying between these faults are upfaulted to form horsts or fault ridges, while they are downfaulted to form graben or fault troughs. The terms "horst" and "graben" are generally applied to structures developed on a regional rather than a local scale. Note that graben is used in both the singular and

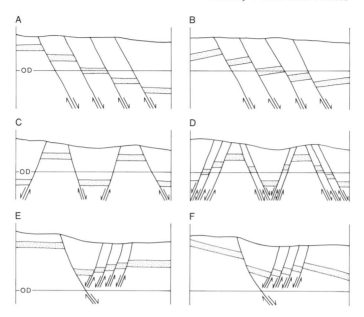

FIG. 5.29. Vertical cross-section through step and block faults: A: Step faults. B: Step faults with tilted fault-blocks. C: Block faults. D: Combination of step and block faults. E and F: Development of antithetic faults (dipping to the left) adjacent to a major fault (dipping to the right).

the plural sense. There is generally little or no tilting of the sedimentary rocks lying between block faults, although the strata may be folded to a certain extent.

Most patterns shown by normal faults result from a combination of step faulting and block faulting. For example, the major faults defining the horsts and graben developed in response to block faulting may be formed by a series of minor step faults. More complex patterns are commonly developed.

It is also found that conjugate sets of normal faults are often developed on different scales, as shown in Figure 5.29E and F. One set may form one or more major faults with large displacements, while the other set forms minor faults with small displacements, which affect the intervening fault blocks. Such faults are known as antithetic faults if they have the opposite sense of displacements to the major faults. Synthetic faults are developed by minor faults with the same sense of displacement, and belonging to the same set of shear fractures, as the major faults.

Normal Faulting in Folded Rocks.

Some patterns shown by normal faults are clearly related to the folding of sedimentary rocks. Longitudinal faults are often developed along the crests of anticlinal folds, as shown in Figure 5.10A. A simple fault trough is formed if the longitudinal faults consist of a pair of normal faults which dip towards one another on either side of the fold hinge. Alternatively, the longitudinal faults can be rather irregular in their development, giving rise to a more complex fault zone which only has the overall form of a fault trough. It is commonly found that these faults, branch and splay towards the plunging ends of periclinal folds.

Normal faulting may also occur at right angles to the fold hinge, as shown in Figure 5.10B. This gives rise to conjugate sets of transverse faults which, downthrowing in either direction, divide the fold into a series of fault blocks. Such faults are often developed at the plunging ends of periclinal folds, although they may occur elsewhere along the fold hinge.

Longitudinal faults are formed by extension affecting the strata as they are folded, as shown in Figure 5.30. This occurs as the result of stretching in the outer arcs of neutral-surface folds. Since this is most likely to occur near the earth's surface, longitudinal faults tend to be developed in anticlines rather than synclines. Transverse faults appear to result from the stretching of the fold hinge as it is arched to form culminations and depressions along the fold trend. Such faults would be equally well developed in anticlines and synclines, as appears to be the case.

Faulting may also result in the development of oblique faults at an angle to the fold hinges. These faults are commonly restricted to the fold limbs, dying out as they are traced in either direction. Although some may be normal faults while others are strike-slip faults, the majority are probably affected by oblique-slip movements to a greater or lesser extent.

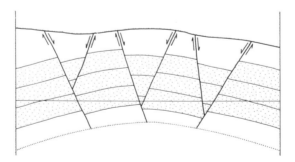

Fig. 5.30. Vertical cross-section through a fold showing the development of normal faults as the result of neutral-surface folding.

En Echelon Normal Faults.

Normal faults may also be developed *en echelon* as the result of strike-slip faulting in the basement rocks underlying a flat-lying sedimentary sequence. This zone of normal faults follows the trend of the underlying strike-slip fault, while the individual faults are aranged at an angle of approximately 45° to this direction, as shown in Figure 5.31. They are formed in response to extension across the fault zone, giving rise to a series of fault blocks which are bounded by conjugate sets of normal faults.

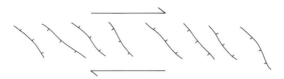

FIG. 5.31. Map of en echelon normal fault developed in response to right-lateral strike-slip movements in the underlying basement.

Mid-oceanic Ridges and Continental Rift Valleys.

Finally, it can be noted that a pattern of normal faulting on a world-wide scale is associated with the formation of mid-oceanic ridges and continental rift valleys. Although the structure of the mid-oceanic ridges is not well known according to the standards of classical geology, the continental rift-valleys are formed by graben, defined by normal faults dipping towards one another and separated by horsts and tilted fault blocks. These rift valleys occur as bifurcating systems along the crests of broad domes which affect the earth's surface. The width of individual rift valleys varies within the range of 30-70 km, corresponding approximately to the thickness of the earth's crust, while the displacements on the normal faults forming their margins can amount to more than 5 km. The fault blocks on either side of these rift valleys are usually tilted slightly so that the highest elevations occur along the crests of the fault scarps. These rift valleys are the result of a long history of earth movements, marked by the accumulation of sediments within the graben and accompanied by various episodes of volcanic activity. A series of half-graben may be formed if only one set of normal faults is developed.

Although a compressional origin has been entertained in the past, it is now accepted that rift valleys are the result of horizontal tension. This develops during the doming and stretching of the earth's crust to form the broad swells now traversed by the rift valleys. The evidence of scale-model experiments shows that similar structures are developed in a clay cake subjected to horizontal tension by the inflation of a rubber balloon.

Upthrusts or High-angle Reverse Faults

Upthrusts are generally formed by single sets of high-angle dip-slip faults on which reverse displacements have taken place. This means that the hanging wall has moved upwards relative to the foot wall, as shown in Figure

5.32. Such faults may be part of a fault system formed by low-angle over-thrusts at depth, which commonly steepen towards the surface. However, we shall only consider in this section those upthrusts which are not related to low-angle faults at depth.

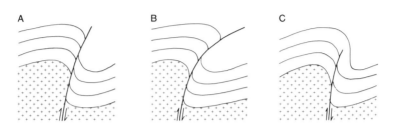

FIG. 5.32. Vertical cross-sections through upthrusts: A: Simple upthrust. B: Upthrust passing upwards into a low-angle overthrust. C: Upthrust passing upwards into a monocline.

As defined, upthrusts form in response to differential uplift and sub-sidence. They are typically developed as boundary faults which form the margins of uplifted blocks. Such blocks are generally composed of igneous and metamorphic rocks underlying a sedimentary sequence and forming a basement of older rocks. The depth of erosion may be such that the sedimentary rocks are found faulted against the basement rocks. However, depending on the depth of erosion, the upthrusts may only be seen to affect the sedimentary rocks at a high level, or the basement rocks at a low level. It is found that some upthrusts become less steep as they are traced upwards from near-vertical faults in the basement to low-angle overthrusts in the overlying sediments. This may be partly due to the lateral spreading of the uplifted block under the influence of gravity. However, theoretical con-siderations suggest that upthrusts would become less steeply inclined as they approach the surface. This tendency might well be accentuated by lateral spreading. Upthrusts affecting basement rocks often change into monoclinal flexures as they pass upwards into the overlying sediments.

Distinguishing Features.

It must be admitted that upthrusts are difficult to distinguish from normal faults unless the dip of the fault plane can be determined. Thus, the fault lines are rarely straight, and they may pass into monoclinal flexures in the same way as normal faults. However, upthrusts are more often associated with folded sequences of sedimentary rocks. The folding com-monly takes the form of monoclinal flexures and over-turned folds, which are somewhat irregular in their development. The upthrusts are generally found to occur on the steep and overturned limbs of these folds. This means that upthrusts, in comparison with normal faults, appear to have a much greater effect on the surrounding rocks, giving rise to steeply dipping and overturned strata and showing a close relationship to the folds so developed.

Low-angle Faults

We have already considered the problems concerning the terminology of low-angle faults. Theoretical considerations indicate that such faults are most likely to form as conjugate sets of low-angle reverse faults, dipping in opposite directions at approximately 30°, as shown in Figure 5.33. Such a fault system would be developed in response to horizontal compression. However, it is commonly found that only a single set of faults is produced, dipping in the same direction and showing the same sense of reverse displacement. This means that differential movements have taken place, resulting in the relative displacement of the upper layers in a particular direction.

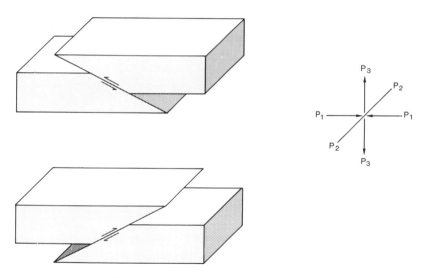

Fig 5.33. Conjugate nature of thrust-faulting.

Break Thrusts.

Such faults are commonly found on the steep limbs of asymmetrical anticlines, as shown in Figure 5.34A. They probably develop in this position as a result of parallel folding, which cannot provide enough space to accommodate the layers in the core of the fold. They occur in response to the folding, so that the fold is the primary structure while the fault is only a

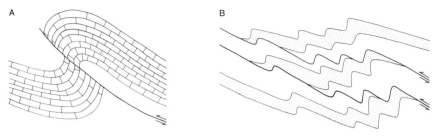

FIG. 5.34. A. Break thrust. B: Low-angle reverse faults.

secondary feature. They differ in this respect from high-angle reverse faults, which cause monoclinal flexures to develop in the rocks affected by the up-thrusting rather than vice versa.

Low-angle Reverse Faults.

Although the thrusts just described are only a local feature, low-angle reverse faults may also be developed on a larger scale. Such faults generally occur as single sets of thrust planes which cut the bedding at a low angle, as shown in Figure 5.34B. They are typically developed in sedimentary sequences which lack marked differences in lithology between the layers. Thrusts of this sort can only develop at an early stage of the deformation, when the strata are still flat-lying. They are commonly folded as the deformation proceeds, giving rise to difficulties in terminology according to the dip of the thrust-plane.

Bedding-plane Thrusts.

Low-angle faults are also developed parallel to the bedding of sedimentary rocks, as shown in Figure 5.35. Such faults can often be traced for long distances, so that they tend to be developed on a very large scale. They are commonly found to affect sedimentary sequences which show marked differences in lithology between the layers. The thrust planes generally follow the incompetent layers in such sequences. Since they are parallel to the bedding, such faults are known as bedding-plane thrusts. They rarely occur as single faults. Instead, several faults occur at different horizons, giving rise to a thrust zone. These structures are formed in response to relative movements of the upper layers in a particular direction, which is generally at right angles to the regional trend.

Thrust zones are formed by a series of thrust sheets or nappes, separated from one another by thrust planes. The thrust sheet lying above a particular thrust plane is sometimes known as the upper plate. The use of this term appears to imply that the thrust sheet acted as a rigid mass during the faulting, although this is rarely the case. The line of a thrust plane is com-

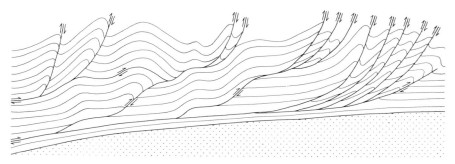

Fig. 5.35. Vertical cross-section through a thrust zone, with the vertical scale much exaggerated, showing the development of bedding plane, step and back-limb thrusts.

monly marked on a geological map by a serrated ornament pointing towards the overlying rocks.

The rocks forming a thrust sheet between two major thrust planes are sometimes cut by a closely spaced set of high-angle reverse faults, as shown in Figure 5.36. These minor faults dip more steeply than bedding in the opposite direction to the thrusts movements. The result is known as imbricate or schuppen structure. The steeply dipping thrust slices resemble the overlapping tiles on a roof.

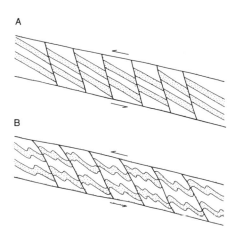

Fig. 5.36. Vertical cross-sections showing imbricate structure between two thrust planes.

The rocks forming thrust sheets are said to be allochthonous, since they are no longer where they were formed. These rocks are known as the allochthon. They have been thrust into their present position from elsewhere. They are separated from autochthonous rocks lying below the allochthon by a sole thrust, formed by the thrust plane at the base of the thrust mass. The autochthonous rocks are still in place. They form the autochthon. The thrust movements are directed towards an area of autochthonous rocks, lying beyond the line of the sole thrust, which is known as the foreland. It may be possible to trace the allochthonous rocks in the opposite direction across the thrust zone into an area known as the root zone, where they would be autochthonous. All these terms imply that overthrusting rather than under-thrusting has taken place.

Step Thrusts.

Although individual thrust planes may follow the bedding for long distances, it is commonly found that they have a step-like character. Such faults cut across the stratigraphic sequence by stepping upwards at an oblique angle from one horizon to another as they are traced towards the foreland. They are best known as step thrusts. The upward stepping of such a low-angle fault may eventually allow the thrust plane to emerge at the earth's surface, forming a surface or erosion thrust. The thrusting causes a sheet of allochthonous rock to advance across the earth's surface, incorporating erosional products such as conglomerates into the fault breccia.

Folding in a Thrust Zone.

Folding as well as faulting often affects the rocks of a thrust zone. The thrust planes can act as surface of décollement, which allow parallel folds to develop in the competent layers forming the thrust sheets. Anticlinal folds are commonly associated with step thrusts, since each step acts as a bulwark against which the rocks can be folded. Splay faults are often developed on the back-limbs of these anticlines, giving rise to back-limb thrusts. Structures of this sort are usually found near the front of a thrust zone, where the thrusts curve upwards towards the surface. Thrust slices are formed by these high-angle reverse faults. It is also the case that folding can occur after the thrusting. Folds are then produced which affect not only the competent layers forming the thrust sheets but also the thrust planes separating the thrust sheets from one another.

Detachment Faults.

The low-angle faults developed within a thrust zone are generally known as thrusts or overthrusts. It has already been mentioned that the use of this term implies horizontal compression, which results in the thrusting of the allochthon over the autochthon. However, it is possible that such faults develop by gravitational sliding on a surface which is gently inclined in the direction of the foreland. Such faults have been termed detachment faults. This term was originally introduced to describe the low-angle fault formed at the base of an allochthonous mass which broke up into isolated fault blocks as it became detached from its substatum along a bedding plane. The surface where the back of the allochthon broke away from the rocks which remained in place is known as the breakaway. If the allochthonous mass moved as a whole, there would be a gap between the breakaway surface and the rear of this mass. This gap is said to be formed by "tectonic erosion". Gravitational sliding rather than overthrusting must have occurred if such a gap can be recognised, or if it can be shown that the allochthonous mass became separated into isolated blocks which moved independently of one another. Such evidence would preclude the possibility that the allochthonous mass was thrust towards the foreland by forces transmitted from its rear.

It should be noted that the use of the term "detachment fault" to describe low-angle faults formed by gravitational sliding has been broadened to include low-angle faults in general. However, this usage is open to criticism since it was originally introduced as a genetic term implying a particular mode of origin which often cannot be substantiated.

Characteristic Features.

Low-angle faults can generally be recognised on a geological map because their outcrop is clearly affected by the topography. They are commonly strike faults, so that the fault lines are often parallel to the boundaries of the stratigraphic formations. Low-angle faults are usually formed by older rocks resting on top of younger rocks. Commonly, the allochthon is formed by the igneous and metamorphic rocks of a basement, lying above a younger sequence of sedimentary rocks. The allochothonous rocks are often more deformed and metamorphosed than the autochothonous rocks, even although they may form part of the same sedimentary sequence. The two groups of rocks may also show marked differences in sedimentary facies, suggesting that they were originally deposited at some distance from one another.

Klippen and Windows.

Special terms are introduced to describe the outcrop pattern of a thrust sheet, as shown in Figure 5.37. An erosional remnant of a thrust sheet, isolated from the main outcrop, is known as a klippe (klippen). The rocks forming a klippe rest on a thrust plane. They are surrounded on all sides by rocks belonging to either an underlying thrust sheet or the foreland. The opposite relationships are seen in a window or fenster. This is an area where a particular thrust sheet is breached by erosion, thereby exposing the rocks which lie beneath. These rocks may belong to an underlying thrust sheet or

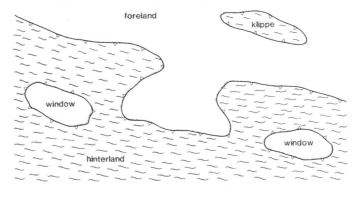

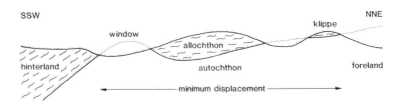

Fɪɢ. 5.37. Map and vertical cross-section showing outcrop pattern typically developed by low-angle thrust faults.

they may represent the autochthon. They are surrounded on all sides by the allochthonous rocks of the overlying thrust sheet.

These features of the outcrop pattern may be developed in a flat-lying thrust sheet as a result of the topography. Klippen would be formed by the higher ground, while windows would tend to occupy the lower ground. Alternatively, if the thrust sheet is folded, klippen are formed by synforms and basins while windows are associated with antiforms and domes. Although klippen and windows are analogous to stratigraphic outliers and inliers, respectively, a clear distinction should be made between the two sets of terms.

The minimum width of an overthrust may be measured from its rear in a window to its front in a klippe, at right angles to the regional trend. This distance may be a hundred kilometres or so, although it is usually less. However, it should be appreciated that this figure does not correspond to the displacement on the overthrust. In fact, the displacement may be more or less than the minimum width of the overthrust. It can only be determined from the overlap of rocks which are recognised to be equivalent to one another above and below the thrust plane. A minimum value is given by the stratigraphic separation, which corresponds to the stratigraphic thickness of the rocks cut out by the thrust plane.

Strike-slip Faults

Ideally, strike-slip faults form conjugate sets of vertical fractures on which strike-slip movements have taken place. These fractures intersect one another in a vertical direction, so that they strike at an oblique angle to one another. Such a fault system develops in response to a horizontal compression, combined with a horizontal extension at right angles to this direction. The acute angle between the two sets of strike-slip faults faces the direction of maximum compression, as shown in Figure 5.38.

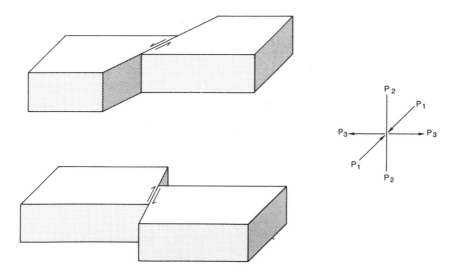

FIG. 5.38. Conjugate nature of strike-slip faulting.

Since strike-slip faults are the result of horizontal compression, they are commonly found to be associated with other structures, such as overthrusts and folds, which develop under similar conditions. They differ from these structures, however, in that the extension accompanying their formation is sidewards rather than upwards. This may reflect the arcuate nature of many thrust zones and fold belts, which causes the rocks to be stretched sidewards as they are compressed by the thrusting and folding against the foreland.

Although strike-slip faults often occur as conjugate sets, it is common for one set to be developed in preference to the other. The faults belonging to the predominant set are more abundant, and usually show greater displacements, than the faults of the subsidiary set. Alternatively, only a single set of strike-slip faults may be developed, sharing a common strike and sense of displacement, equivalent to one of the conjugate sets.

Strike-slip faults have been termed "wrench faults". This term implies that rotational movements have taken place in a horizontal plane. Its use is only appropriate where the conjugate sets of strike-slip faults are not developed to the same extent, or where only a single set of strike-slip faults is present.

Strike-slip Faults in Thrust Zones and Fold Belts.

Strike-slip faults are often associated with the development of overthrusts in thrust zones, as shown in Figure 5.39A. Since they form in response to a horizontal compression, which takes place in the direction of overthrusting, they generally have oblique trends at a high angle to the thrust zone itself. Such faults may divide the thrust sheet into thrust blocks which advance independently of one another on the underlying thrust plane. Such faults have been termed tear faults. Alternatively, strike-slip faults may be developed so that they change along their course into overthrusts. This can only occur if the fault plane is curved to form a cylindrical surface. This has an axis which is parallel to a common direction of displacement on the strike-slip fault and the overthrust. Accordingly, there would be a component of dip-slip movement on the strike-slip faults, depending on the dip of the thrust plane. It is sometimes found that over-thrusts are joined together by strike-slip faults of this sort, giving rise to sigmoidal deflections in the trend of the fault lines.

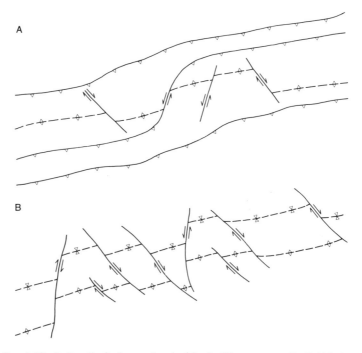

Fig. 5.39. Strike-slip fault associated with: A: Thrust zones. B: Fold belts.

Folds belts are another setting where strike-slip faults may be developed. They generally occur as oblique faults trending at an angle of approximately 60° to the axial traces of the folds, as shown in Figure 5.39B. The displacement on these faults can usually be determined from the offset shown by the axial traces of the individual folds as they cross the fault lines.

It is sometimes found that these folds show marked differences in style as they are traced from one fault-block to another. This suggests that the faulting took place at the same time as the folding. If the folding occurs as the result of décollement along an underlying thrust plane, the strike-slip faults affecting the folded rocks are known as tear faults. The displacement on these faults occurs in order to accommodate the differences which result from the folding of the fault blocks on either side. Tear faults commonly merge into overthrusts as they are traced along their strike.

Strike-slip faults may also be formed in fold belts after the folding has taken place. The folds can then be traced across the fault lines without showing a marked change in style. These faults are commonly developed in a more regular fashion, so that they form a fault pattern which is not closely related to the presence of individual folds or overthrusts.

Regional Strike-slip Faults.

Strike-slip faults may also be developed on a regional scale, without any connection with folding or thrusting, as shown in Figure 5.40. It is commonly found that only one major fault is developed in this case, with a

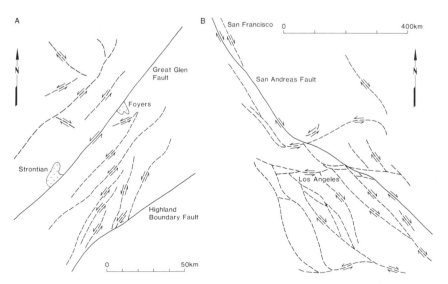

FIG. 5.40. Map of regional strike-slip fault systems: A: Great Glen Fault (Scotland). B: San Andrens Fault (California).

displacement of a hundred kilometres or more. However, such a fault is usually accompanied by a number of minor faults, on which the displacements are much less. Although most of these faults belong to the same set as the major fault, it is sometimes found that the conjugate set is also developed. The displacements on these minor faults can often be determined by matching of the structures on either side of the fault line. This becomes increasingly difficult as the displacement becomes greater. Eventually, it is necessary to use features occurring on a regional scale in order to determine the displacements.

The displacement of sedimentary facies, granite plutons, zones of high grade metamorphism, and tectonic zones have all been used to determine the movements on such strike-slip faults as the San Andreas Fault (California), the Great Glen Fault (Scotland) and the Alpine Fault (New Zealand). It is important to realise that the displacements on major (and minor) faults are cumulative in that they have developed over a long period of geological time through a series of small movements. Thus, it is only the oldest features which allow the total displacement to be determined.

Transform Faults.

The final type of strike-slip fault to be considered are transform faults. These form the transverse fracture zones which are part of the world-wide system of mid-oceanic ridges developed as a result of sea-floor spreading. Intrusion of magma occurs below the mid-oceanic ridges, causing the rocks forming the sea-floor on either side to move apart and resulting in the formation of new oceanic crust, as shown in Figure 5.41. The ridge system formed in this way consists of short segments of mid-oceanic ridge, linked together by transverse fracture zones. The mid-oceanic ridge sidesteps left or right as it crosses these fracture zones, which are generally arranged at right angles to the trend of the ridge itself.

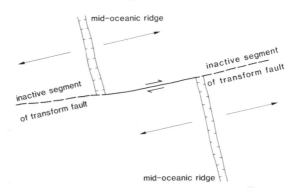

FIG. 5.41. Map of a transform fault associated with a mid-oceanic ridge system.

The formation of oceanic crust at the mid-oceanic ridge causes displacements to occur along the fracture zones. Inspection of Figure 5.41 shows that the sense of this displacement is right-lateral if the mid-oceanic ridge sidesteps to the left, and vice versa. Moreover, these displacements only occur along that part of the fracture zone linking the ends of the mid-oceanic ridge segments. The fracture zone is inactive where it extends beyond the mid-oceanic ridge system towards the continents on either side. Movements only took place on this part of the fracture zone at an earlier stage of its evolution.

The strike-slip faults formed in this way are sufficiently distinct to be classed separately as transform faults. However, it is quite likely that the major strike-slip faults described in a previous paragraph represent a particular type of transform fault. These major faults appear to be formed by transform faults which cut across continental areas from one mid-oceanic ridge system to another. The process of sea-floor spreading would then allow the large displacements to occur on these strike-slip faults, which are otherwise difficult to accommodate at the ends of the fault systems. The San Andreas Fault (California) and the Alpine Fault (New Zealand) provide clear examples of such a connection between major strike-slip faults in continental areas, and the mid-oceanic ridge systems.

Characteristic Features.

Strike-slip faults can be identified from a geological map if there are corresponding features on either side of the fault line which allow the displacement to be determined. It has already been emphasised that the strike separation shown by a geological contact does not provide sufficient evidence to do this. It is often the case that the features required to determine the displacement on a strike-slip fault are lacking. This means that strike-slip faults can often only be recognised from the general nature of the fault pattern.

Since the fault planes are usually vertical or nearly so, the topography has little or no effect on the fault lines. However, strike-slip faulting is often accompanied by the widespread brecciation and cataclasis of the country rocks. This produces wide fault zones which are often attacked by differential erosion to form conspicuous fault-line valleys.

It is also characteristic feature of strike-slip faults that the fault lines are straight and rectilinear. They differ in this respect from normal faults and upthrusts which tend to show marked changes in trend along their course. However, strike-slip faults associated with folds and overthrusts are often curved towards their extremities, where they may merge with other structures.

The pattern formed by conjugate sets of strike-slip faults may also be diagnostic, since it gives rise to faults trending at an oblique angle to one another. The displacements on these faults should be directed towards the acute angle between the fault sets. However, similar patterns are shown by intersecting sets of normal faults.

Fault Rocks

Although some faults occur as clean-cut fractures, most faults are accompanied by the disruption and mechanical breakdown of the wall rocks on either side of the fault zone. This is marked by the fracturing, crushing and granulation of the wallrocks, and results in the formation of what are known as fault rocks. The mechanical breakdown of the wallrocks in response to such processes is generally known as cataclasis, while the fault rocks so formed may be termed cataclastic rocks.

Fault Breccias and Fault Gouges.

Fault breccia is a common type of fault rock produced by the fracturing of brittle wall rocks to form a jumbled mass of angular rock fragments which may vary considerably in size. These fragments are sometimes embedded in a finer-grained matrix of crushed and pulverised material. Alternatively, they may be cemented together by the precipitation of vein minerals such as quartz, calcite, dolomite or barytes in the cavities which occur between the fragments. Some faults are formed by discrete zones of fault breccia showing a sharp contact with the wall rocks on either side, while other faults show a more gradual transition, marked by a decrease in the amount of brecciation affecting the wall rocks away from the fault zone. The angular nature of the rock fragments forming a fault breccia means that the wallrocks were affected by fracturing in a very irregular manner. However, there is a complete transition from fault zones lined with fault breccia to sheeted or braided fault zones formed by anastomasing fractures which are more-or-less parallel to the walls of the faults.

Fault breccias should be clearly distinguished from crush breccias and crush conglomerates which are formed in place, or nearly so, by the crushing of closely jointed layers of brittle rock. These are formed by lozenge-shaped fragments separated from one anther by intersecting or anastomasing sets of shear fractures. The fragments in crush breccias are more angular and less rounded than the fragments in crush conglomerates. They are often isolated from one another by a matrix of finer-grained material, giving the appearance of sedimentary breccias and conglomerates. Indeed, many rocks described as crush breccias or crush conglomerates in the literature appear to be deformed breccias and conglomerates produced originally by the sliding or slumping of partly consolidated sediments.

Fault breccia does not form if the wallrocks have a lithology which favours mechanical breakdown by crushing and grinding to form a finely pulverised powder with the consistency of clay. This produces a fault rock known as fault gouge.

Fault breccia and fault gouge are rocks which form near the surface where the confining pressure due to the weight of the overlying rocks is relatively low. A characteristic feature of such rocks is that they are not indurated or lithified unless the broken fragments forming the rock have later been cemented together into a cohesive mass. They differ in this respect from fault rocks such as microbreccia, mylonite and ultramylonite which are formed as indurated or lithified rocks. requiring no later cementation to develop a cohesive character. Such rocks are produced under deep-seated conditions where the confining pressure is relatively high. They are all fine-grained or cryptocrystalline rocks formed by the intense fracturing and granulation associated with fault movements.

Microbreccias, Mylonites and Ultramylonites.

Microbreccias are intensely fractured rocks consisting of angular fragments lying in a matrix of finer-grained material. Such rocks are termed cataclasites if most of the fragments are less than 0.2 mm in diameter and comprise less than 30 per cent of the rock. Microbreccias and cataclasites lack any fluxion structure which is formed by the grinding and commi-

TABLE 5.1

Classification of Cataclastic Rocks (after Higgins, 1971)

		Cataclasis	dominant	Recrystallisation dominant
		No fluxion structure	Fluxion structure	Fluxion structure
Percentage volume of rock fragments of porphyroclasts	50%	Microbreccia > 0.2 mm	Protomylonite > 0.5 mm	Mylonite schist or Mylonite Gneiss > 0.5 mm
	30%	Cataclasite < 0.2mm	Mylonite > 0.2 mm	
	10%		Ultramylonite < 0.2mm	Blastomylonite < 0.5 mm

Note: Figures in millimetres give approximate grain-size limits for rock fragments or Porphyroclasts in different rock types.

nution of fragments in the rock along shear planes. They differ in this respect from mylonites and related rocks.

Mylonites are cataclastic rocks which show fluxion structure in outcrop or under the microscope. The initial stages in the formation of a mylonite are marked by the development of a rock known as protomylonite. This consists of lenticular fragments of the original rock, separated from one another by anastomasing films of fine-grained material which is produced by cataclasis along shear planes. These shear planes often follow the grain boundaries in monomineralic rocks such as quartzite, so that the finely ground material has the same composition as the fragments lying between the shear planes. However, if cataclasis affects rocks formed by an assemblage of different minerals such as granite, it is commonly found that these minerals differ in their resistance to mechanical breakdown. For example, quartz generally offers less resistance to cataclasis than felspar, so that it is the quartz grains which tend to be reduced by granulation along the shear planes, leaving the felspar grains isolated as lenticular fragments.

Mylonite is produced from protomylonite as the result of further movement. The lenticular fragments found in protomylonite are reduced in size, and become rounded, in response to the continued grinding and milling which occurs along the shear planes. These fragments are generally more than 0.2 mm in diameter and comprise 10 to 50 per cent of the rock. They are known as porphyroclasts. A streaky, banded or platy flow structure known as fluxion structure is developed in the finely comminuted matrix as a result of the differential movements which have affected the rock. Although this may be expressed as a colour lamination or a compositional banding which can be seen in the field, the definition of mylonite only requires that some sort of fluxion structure can be recognised under the microscope.

The final stage in the formation of mylonite rocks as the result of cataclasis is marked by the development of ultramylonite. This is a cryptocrystalline rock produced by extreme comminution. Porphyroclasts are generally less than 0.2 mm in diameter and comprise less than 10 per cent of the rock. Many of these porphyroclasts have been reduced to streaks of microbreccia. Ultramylonites are dense and compact rocks which show fluxion structure, at least under the microscope.

Pseudotachylites and Flinty Crush-rocks.

The frictional heating associated with fault movements may be sufficient to melt cataclastic rocks in fault zones. This possibility is enhanced if the movements affect rocks, such as igneous intrusions and their aureoles, which were hot at the time. Such melts may intrude the fault zone and its wallrocks to form dykes, veins and stringers of a rock known as pseudotachylite. This is a dark glassy rock resembling basaltic glass (tachylite). The glassy matrix may show evidence of incipient crystallisation, while amygdales and vesicles can be produced by the release of gas. Rock and mineral fragments enclosed by pseudotachylite may be affected by partial melting. If clear evidence of complete melting is lacking, such rocks are more properly termed flinty crush-rocks. This name has been used in the past as a field-term for a variety of fine-grained or cryptocrystalline, cataclastic rocks with a flinty appearance. It is best applied to cataclastic rocks which show intrusive relationships while lacking conclusive evidence that they were ever molten. However, it should be noted that some flinty crush-rocks are formed by the extreme comminution of rock and mineral fragments entrained by the flow of volcanic gas.

Augen Gneisses and Blastomylonites.

Cataclasis is the dominant process involved in the formation of all fault rocks so far considered. While mylonitic rocks are thought to form under conditions of high confining pressure, it is likely that the cataclastic processes of mechanical breakdown are favoured by low temperatures. If the movements occur at higher temperatures, if is found that recrystallisation and the growth of new minerals becomes increasingly important. These processes are essentially metamorphic, so that cataclastic rocks such as mylonite tend to grade into metamorphic rocks such as schists and gneisses. An intermediate stage is marked by the formation of fault rocks showing evidence of cataclasis and recrystallisation. Such rocks can be classed as low-grade metamorphic rocks produced in response to extreme deformation under conditions of moderate temperature. They consist of porphyroclasts set in a finer-grained matrix which shows evidence of considerable recrystallisation. The porphyroclasts represent relicts of the mineral grains which originally formed the whole rock. They show a gradual reduction in size and number as the movements become more intense.

Such rocks are known variously as mylonite gneisses, mylonite schists and blastomylonites. Most of the prophyroclasts in mylonite gneisses and mylonite schists are more than 0.5 mm diameter, while they comprise at least 30 per cent of the rock. Such rocks are roughly the equivalent of protomylonite except that recrystallisation is dominant over cataclasis. As suggested by the terminology, mylonite gneisses resemble gneisses in that they are coarsely banded rocks with an irregular and discontinuous fabric, while mylonite schists are more like schists in that they are finely foliated rocks with a regular and closely spaced fabric. Since the porphyroclasts in mylonite gneisses form eyes round which the foliation is diverted, such rocks are sometimes known as augen gneisses. An incipient form of this fabric is known as flaser structure where it affects coarse-grained rocks such as granites or gabbros. Blastomylonite is roughly the equivalent of mylonite or ultramylonite except that recrystallisation is dominant over cataclasis.

Most of the prophyroclasts in blastomylonite are less than 0.5 mm in diameter and comprise less than 30 per cent of the rock.

Phyllonites.

Cataclasis and recrystallisation may also affect metamorphic rocks formed originally under conditions of high temperature. The minerals in such high-grade rocks are broken down to form stable assemblages under the conditions of lowered temperature. This reduction of metamorphic grade which accompanies the cataclasis and recrystallisation of high-grade rocks is known as retrograde metamorphism. It is associated with the development of rocks known as phyllonites.

CHAPTER 6

Igneous Rocks and their Structure

Introduction

IGNEOUS rocks are produced by the solidification of magma, which is the name given to molten-rock material. Although some magmas may be formed in place by the melting of solid rocks, most are derived from considerable depths within the earth's crust and, particularly, the upper mantle. Once formed at depth, this material ascends towards the earth's surface under pressure, intruding the pre-existing rocks of the earth's crust to form igneous intrusions. The magma retained within these intrusions solidifies to form the intrusive rocks, surrounded by the older country-rocks of the earth's crust. The remaining magma reaches the earth's surface to erupt as lava from volcanoes, which solidifies to form the extrusive rocks. They occur in the form of lava flows. The eruption of lava flows is often associated with explosive activity, producing fragmental material which accumulates as the pyroclastic rocks. A magma chamber is developed whenever these extrusive rocks appear to be derived from an intrusion within the crust, rather than directly from the source of the magma at a greater depth within the upper mantle.

A liquid silicate-melt at a temperature of 800-1100°C is the essential constituent of most magmas. This melt can vary considerably in composition. However, we need only distinguish two magma types. Acid magmas are rich in silicon, sodium and potassium but poor in iron, magnesium and calcium. They solidify to form the granitic rocks. Basic magmas are relatively poor in silicon, sodium and potassium but rich in iron, magnesium and calcium. They solidify to form the basaltic rocks. Varying amounts of volatile substances are generally dissolved in magmas, to be released as gases and vapours when it solidifies. Water vapour and carbon dioxide are the most common of these constituents. Solid crystals and rock fragments may also be carried in suspension by the magmas. The rock fragments are preserved as xenoliths in the igneous rock, once the magma has solidified, as shown in Figure 6.1.

How easily a magma can flow depends on its viscosity. This varies according to the composition, temperature and volatile content of the magma. In general, basic magmas are relatively fluid, while acid magmas are highly viscous. The presence of volatile substances in solution lowers the

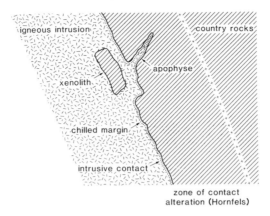

FIG. 6.1. Diagram showing some of the features associated with the contact of an igneous intrusion.

viscosity. However, magmas become more viscous as they cool towards the temperature at which they start to solidify. Indeed, rapid cooling can so increase the viscosity that the magma is converted into a supercooled glass. This can occur if the magma intrudes cold country rocks or when it comes into contact with air or water at the earth's surface. Otherwise, a magma simply solidifies by crystallisation. The grain size of the igneous rocks formed in this way depends on how quickly the magma cools. Rapid cooling allows many small crystals to form, so that a fine-grained rock is produced, whereas slow cooling has the opposite effect, allowing a coarse-grained rock to be formed.

How quickly a magma loses heat to its surroundings depends on the relative size of the intrusion. The small size of minor intrusions means that they cool rapidly, so that they are formed by fine-grained rocks, whereas the large size of major intrusions means that they cool slowly, so that they are formed by coarse-grained rocks. Moreover, since heat is lost outwards to the country-rocks, an igneous intrusion cools from its margins inwards. This allows fine-grained or glassy rocks to be formed along a chilled margin with the country-rocks, even although coarser-grained rocks form the interior parts of the intrusion.

The country rocks surrounding an igneous intrusion may be so heated by the magma that they are affected by contact or thermal metamorphism. Typically, an aureole of contact metamorphism is developed around the igneous intrusion, within which the country rocks are converted by recrystallisation into a metamorphic rock known in general as a hornfels. Such aureoles of thermal metmorphism are generally only found around major intrusions, even although there may be evidence of thermal alteration in the immediate vicinity of minor intrusions. The limits of such contact metamorphism are often shown on geological maps by means of a special ornament.

Lava Flows

Magma erupts from volcanoes to form lava flows. Their form depends to a considerable extent of the viscosity of the magma. Thus, fluid magma tends to flow for long distances across the earth's surface, taking up the form of this surface as it does so. If this surface is relatively flat, the lava would form a thin lava-flow of wide extent. However, if the eruption takes place in an area of greater relief, the lava tends to flow downhill into the valleys, where it forms thicker flows confined to the lower ground by the valley sides. The repeated eruption of fluid lava would gradually fill in these valleys, until a flat-lying lava field was formed. Subsequent eruptions would then produce thin lava-flows extending widely over this lava field. Such a tabular or sheet-like form can be considered as typical of fluid lava-flows. By way of contrast, viscous lava does not flow for any distance from its source before it solidifies. Such lava tends to form thick lava-flows of limited extent, unless it is so viscous that it forms a steep-sided dome above the volcanic orifice. The topographic relief has hardly any influence on the form of such a viscous lava flow. Since the viscosity of lava varies according to its composition, temperature and volatile content, a complete gradation in form may be expected between sheet-like lava-flows and steep-sided lava-domes.

Vesicles and Amygdales.

The volatile constituents of a magma, and the ease with which they can escape as a gas phase, exert an important influence on the internal structures of lava flows. Indeed, it is partly the presence of such substances, dissolved in the lava, which allows it to flow with relative ease across the earth's surface. This factor therefore needs to be taken into account when considering the varied form of lava flows. However, the volatile constituents tend to escape as a gas phase as the lava cools and solidifies. This results in the formation of gas bubbles, which may be preserved as vesicles if the gas does not reach the surface before the lava-flow solidifies. These cavities vary considerably in shape, since they may be spherical, ellipsoidal, cylindrical or irregular in form. They are commonly filled with mineral matter, so forming amygdales. Since any gas bubbles in the lava would tend to rise towards its surface, vesicular and amygdaloidal structures are typically developed in the upper levels of the flow. Indeed, the tops of lava-flows are often marked by the development of highly vesicular or even scoriaceous rocks, distended by the presence of abundant and irregular cavities of all sizes to form a clinker-like mass. The contrast between the upper levels of a lava-flow, charged with vesicles, and its more massive interior is commonly marked in the topography by the development of trap features. The interior parts of the individual flows resist weathering and erosion, so that they form scarps and steeper slopes, while the upper levels of the individual flows are attacked preferentially to form flat-lying features in the topography.

Ropy and Blocky Lava.

How easily the volatile constituents can escape from the lava depends to a considerable extent on its temperature and composition. If the lava is relatively hot, gas can escape with relative ease from the surface layers, which tend to solidify as a glassy and highly viscous skin. The continued flow of the underlying lava contorts this smooth surface into wrinkles, corrugations and folds as it is carried along, so that it exhibits the form of what is known as ropy or pahoehoe lava. However, if the lava is not so hot, gas cannot escape so easily. The surface layers then tend to form a partly crystallised crust, which breaks up as the underlying lava continues to flow. By reducing the pressure on this lava, the volatile consituents are released in sudden bursts as it continues to crystallise, so forming what is known as blocky or aa lava. This consists of a jumble of highly vesicular blocks, carried along by the flow of the underlying lava. Although such a blocky structure is only likely to be developed at the upper surface of a fluid lava, it can affect the greater part of a more viscous flow, to form what has been termed a flow breccia. This typically consists of angular blocks of relatively coarse-grained and vesicular lava, separated from one another by irregular masses of finer-grained and less vesicular lava, representing the last portion of the lava-flow to solidify.

Flow Structures.

Apart from the presence of vesicular and amygdaloidal structures, and the development of flow breccias, lava-flows mostly lack any internal structure. However, rather than forming massive rocks, as commonly the case, a flow lamination may be developed. This is typically the result of the parallel alignment of platy and acicular minerals caused by the lava continuing to flow as it crystallised. Vesicles and amygdales are commonly elongated in the direction of flow within this lamination. It grades into a flow banding wherever compositional differences are developed between different layers in the rocks. It is commonly found that flow banding becomes complexly folded in the more viscous flows of acid composition, such as rhyolites.

Evidence of Stratigraphic Order.

Lava-flows erupted on land commonly show a number of features which can be used to distinguish the top from the bottom, as shown in Figure 6.2A. The lower part of a lava-flow tends to chill against the underlying rocks, so that it may be formed by a fine-grained or glassy rock. If the lava has flowed across soft ground, as might be formed by muddy sediment, gigantic load casts would be formed in its base, with tongues of the underlying sediment penetrating upwards into the lava-flow between bulbous masses of lava. The gases released from the lava tend to escape upwards as gas bubbles, which accumulate in the upper part of the lava-flow as vesicles and amygdales. However, amygdales or vesicles in the form of pipes may also be found near the base of a lava-flow. They often coalesce upwards as two adjacent amygdales join together to form an inverted Y-shape. The release of gas may also cause the upper part of the lava flow to break-up into a jumble of highly vesicular blocks, known as scoria. This means that the top of a lava-flow can commonly be distinguished by its vesicular and often scoriaceous character. Even if these features are not developed, the upper surface is usually more irregular than the lower contact. For example, sediment is often found penetrating cracks and cavities in the upper surface of a

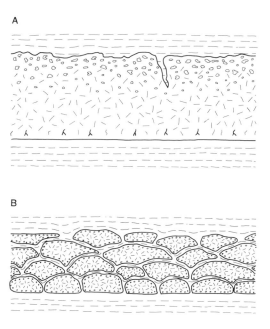

Fɪɢ. 6.2. Evidence of stratigraphic order in lava-flows. A: Chilled lower contact with pipe amygdales contrasts with vesicular upper part of a lava flow. B: Pillow structure in a lava flow erupted under water.

lava-flow. This surface can also be affected by weathering, forming a soil profile which may be preserved if the lava-flow is then covered by subsequent eruptions.

Lava-flows erupted into water or beneath a thick cover of snow or ice show a peculiar type of flow structure which is not only diagnostic of their origin but also allows the stratigraphic order to be determined. This is known as pillow structure. It is formed by fluid lavas of basaltic composition, which come into contact with water as they are erupted, so forming pillow lavas. The rapid chilling causes a glassy skin to form on the lava flow as soon as it issues from the volcanic vent. The escape of lava through cracks in this skin allows bulbous masses of fluid lava to form, which cease to grow once they are encased by a glassy skin. The process is repeated to form a whole mass of lava pillows, resting on top of one another. The pillows at the base of the lava-flow typically form bun-shaped masses with rounded tops and flat bottoms, as shown in Figure 6.2B. This shape becomes modified within the lava flow, where the individual pillows project downwards into the spaces between the underlying pillows while retaining the rounded form of their upper surfaces. The contrast between the cuspate bottoms and the rounded tops of the individual pillows then allows the top of the lava-flow to be distinguished from its bottom. Most pillow lavas are erupted on to the sea-floor. Radiolarian chert or siliceous limestone is commonly found in any spaces which are left between the pillows of such a submarine lava.

Pyroclastic Deposits

Pyroclastic rocks are produced by the volcanic explosions which occur as the volatile constituents are released under pressure from magma at shallow depths. These explosions occur if this gas phase cannot escape easily to the surface. The fragments in these rocks may be derived from the lava itself. Thus, liquid lava may be disrupted to form volcanic bombs, cindery lapilli and glass shards, in order of decreasing size, which accumulate as fragments, along with any crystals which would have been present in the magma as it started to crystallise. Alternatively, the fragments may be formed by the disruption of solid rocks, so that they can represent either the country rocks which originally lined the walls of the volcanic orifice or the lava which has already solidified within this orifice.

Pyroclastic rocks are classified according to the size of the constituent fragments. Agglomerates and breccias contain fragments greater than 4 mm in diameter, which are generally set in a finer-grained matrix. Agglomerates are distinguished from breccias according to whether these fragments are round or angular, respectively. Tuffs are finer-grained pyroclastic rocks, with fragments less than 4 mm in diameter. Although agglomerates and breccias are mostly formed by rock fragments or volcanic bombs, tuffs may be formed by rock fragments, crystal fragments, lapilli or glass shards.

Ash Falls.

This fragmental material is mostly deposited as ash falls. This means that the individual fragments have been blown into the air by the force of the volcanic explosions, which formed the fragments in the first place. Some of this fragmental material falls back into the volcanic vent, where it forms vent agglomerate, as shown in Figure 6.3. It is generally composed of the larger fragments which were produced by the explosive activity. The remainder of the fragmental material is ejected from the volcanic vent, to fall over the surrounding area. Obviously, the force of the volcanic explosions can only blow the larger fragments a short distance before they come to rest, while the smaller fragments would be discharged to a greater distance. Accordingly, agglomerates and breccias tend to be deposited around the volcanic vent, so forming a cinder cone with its sides sloping outwards at the 30 - 40° angle of repose for the constituent fragments. The form of such a cone indicates that it was largely constructed by the avalanching of fragmental material down its sides, once the accumulation of this material exceeded the angle of repose for the constituent fragments. A crater would be developed at its top. The finer material is discharged beyond the confines of this cone to form deposits of tuff and volcanic ashes which mantle the surrounding area. Commonly, bedding is developed in all these deposits in response to fluctuations in the volcanic activity, or its periodic cessation.

The finest material in an ash fall can be transported over very wide areas before it is deposited as a thin but very distinctive layer of volcanic dust as part of a sedimentary sequence. Such a layer, which is commonly altered by sea water to form a bentonite clay, makes an ideal stratigraphic marker if is preserved in the geological record.

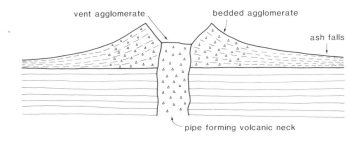

FIG. 6.3. Vertical cross-section through a volcanic vent.

Ash flows.

Pyroclastic rocks can also be deposited in the form of ash flows. Although acid magmas tend to be rich in volatile substances, their high viscosity makes it difficult for these components to be released at any depth as a gas phase. However, this can be achieved if the gas phase comes out of solution in the magma to form gas bubbles, so forming a gas-in-liquid system from the originally liquid magma. The continued release of gas would eventually cause these bubbles to coalesce with one another, so converting the gas-in-liquid system into a liquid-in-gas system, as shown in Figure 6.4. In other words, a foam is converted into a spray. Such a system is capable of flowing as a coherent mass of solid and liquid fragments, suspended by the turbulent motion of hot gas. The solid fragments are formed by any crystals which had already formed in the magma, and by solid rocks which are incorporated into the system by explosive activity. The liquid fragments are formed by clots of gas-rich magma, which solidify to form pumice fragments and glass shards.

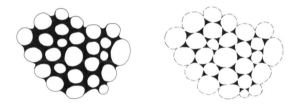

FIG. 6.4. Conversion of a foam into a spray by the coalescence of gas bubbles. Note the cuspate nature of the glass shards (shown in black), as found in ignimbrites.

Once such a system starts to form near the surface, it releases the pressure on the underlying magma. This allows a gas phase to separate from the magma at greater and greater depths, so that the process can sustain itself once it has started. The eruption of this material occurs in the form of a nuée ardente, or glowing avalanche, which issues from the volcanic vent to flow across the earth's surface as a turbulent suspension of solid crystals, rock fragments and glass shards. Nuée ardentes are sufficiently mobile to flow for long distances under the influence of gravity, until the hot mass comes to rest as a widespread sheet of pyroclastic material, which fills in any hollows of the topography. Such a deposit of pyroclastic material is known as an ignimbrite. Sufficient heat is generally retained, which allows the pyroclastic deposit to be welded together under the weight of the overlying material, to form a welded tuff.

Ignimbrites as Stratigraphic Markers.

Ignimbrites commonly erupt over a sufficiently wide areas to act as stratigraphic markers. Indeed, they show several features which are ideal for this purpose. Firstly, ignimbrites erupt so rapidly to form ash-flows that they can be considered to mark a single instant of geological time. This is the case even if more than one ash-flow is formed during a particular phase of eruptive activity, since these ash-flows commonly form several sheets which cool together as a single unit. Secondly, so large a volume of pyroclastic material is produced during each phase of eruptive activity that the underlying chamber becomes drained of magma. Ignimbrite eruptions therefore tend to be occasional events so that there would be only a few horizons to be distinguished from one another in a particular region. Thirdly, the mode of eruption for this large volume of pyroclastic material is such that ignimbrites tend to form thick sheets covering a very wide area. This means that ignimbrite sheets can generally be traced for long distances, unless they are confined within calderas or other depresions. Finally, ignimbrites consolidate to form welded tuff which not only resists the destructive effects of erosion but which is also sufficiently distinct a rock type to be easily recognised.

Evidence of Stratigraphic Order.

Pyroclastic deposits show only a few structures which can be used to determine their stratigraphic order in the field, unless sedimentary agencies have redistributed the fragmental material to form tuffaceous sediments. However, grading of the fragmental material may be developed in ash-fall deposits, particularly if they are water-lain, so that the fragments become larger and denser towards the base of the deposit. Large fragments falling into bedded tuffs may also form indentations below these fragments, breaking through the bedding, which then become covered by the subsequent eruption of pyroclastic material

Lava Plateaux and Central Volcanoes

Volcanoes and lava plateaux are formed by the gradual accumulation of lava-flows, ash-falls and ash-flows around and within a volcanic vent. Their character depends on the nature of the volcanic vent, the fluidity of the lava, and the degree of explosive activity.

Lava Plateaux.

Fissure eruptions take place from a narrow crack or fissure in the earth's crust, which presumably extends downwards to the source of the magma. Such eruptions are favoured by fluid lava of basaltic composition, lacking any explosive tendencies. This lava generally erupts repeatedly from a series of fissures to form thin lava-flows covering a very wide area. Over a period of geological time, a thick pile of lava-flows can be built-up, to form a lava plateau. The name is rather a misnomer, since a lava pile will only form a plateau if it is subsequently affected by erosion and denudation, perhaps as a result of later earth movements.

Central Volcanoes.

The other form of volcanic activity is associated with central eruptions, giving rise to volcanoes which can vary considerably in character according to the nature of the volcanic products. Shield volcanoes are favoured by fluid lavas of basaltic composition, lacking any explosive tendencies. They are formed by thin lava-flows of wide extent, which gradually accumulate to form a very low dome. Strato-volcanoes are built-up by more viscous lava, which has explosive tendencies. They are formed by thick lava-flows of limited extent, interbedded with ash-falls of pyroclastic material, which gradually accumulate in the form of a volcanic cone. Individual volcanoes would be expected to vary in form between shield volcanoes and strato-volcanoes, according to the relative viscosity of the lava being erupted and the degree of explosive activity associated with its eruption. All these volcanoes have a summit crater which forms an opening to the underlying vent. However, there may be subsidiary vents and fissures developed on the sides of the volcano, once it reaches a substantial height above its surroundings.

Calderas.

Central voclanoes may be modified if a caldera is formed as a circular depression of major proportions in the upper part of the volcano, flanked by encircling walls. Although some may be formed by the explosive decapitation of the volcanic superstructure, most calderas are formed by the collapse of this superstructure under its own weight, caused by a lack of support at depth. For example, the copious eruption of lavas from the flanks of a shield volcano can so drain the underlying magma chamber that its roof founders to form a caldera at the surface. Alternatively, the eruption of ignimbrites from a strato-volcano may allow a caldera to develop by subsidence on a series of ring faults. This process is often associated with cauldron subsidence below the earth's surface, leading to the intrusion of ring dykes, as described below.

Statigraphic Record of Volcanic Rocks

Since volcanoes are built up above their surroundings as the result of repeated eruption, they tend to be easily attacked by weathering and erosion. This means that the products of volcanic activity are often so denuded that only the intrusive rocks forming the roots of the volcano are left. However, the load produced by a large volcano may be so great that it depresses the underlying rocks of the earth's crust, thus allowing the lower part of the volcanic edifice to be preserved from denudation. This may also be achieved locally as the result of cauldron subsidence. Otherwise, volcanic rocks are only likely to be preserved in the geological record if their eruption occurs as sedimentation is proceeding, burying the volcanic edifice as it forms. Accordingly, if volcanic rocks are preserved in the geological record, they are commonly found interbedded to some extent with sedimentary rocks, as part of a stratigraphic sequence. In general, such rocks represent either lava plateaux, the lower levels of central volcanoes, or the local effect of cauldron subsidence.

This means in effect that lava-flows and pyroclastic deposits form stratigraphic sequences which become younger in an upward direction, assuming that the rocks have not been overturned as a result of folding or faulting. The Principle of Superposition and the methods used to determine the age and stratigraphic order of sedimentary rocks can therefore be applied with little modification to volcanic rocks. Fossils are not commonly preserved in volcanic sequences, so that the stratigraphic age can usually only be determined by reference to any sedimentary rocks which might be associated with them. However, lava-flows and pyroclastic deposits do show primary structures which can be used to determine the stratigraphic order of the volcanic sequence.

Mapping of Volcanic Rocks

Volcanic rocks are usually distinguished on geological maps by means of a special colour or ornament, while they are assigned symbols similar to those used for sedimentary formations. The symbols used by the United States Geological Survey comprise a capital letter, which gives the stratigraphic age according to the geological systems recognised in the third column of Table 2.1, followed by a short abbreviation of one or more letters, which gives the rock type or the formation name. The symbols generally used by the Geological Survey of Great Britain comprise a short abbreviation starting with a capital letter, which gives the rock-type as shown in Table 6.1, followed by the normal symbol for the lithostratigraphic group, as shown in the fifth column of Table 2.1. For example, Bc^1 is used for basaltic lavas belonging to the Lower Old Red Sandstone. The explanation to the geological map usually states explicitly that the ornament and symbols used in this way apply to extrusive lava-flows and pyroclastic deposits.

Nature of Volcanic Formations.

Volcanic rocks are mapped in the same way as sedimetary formations. On occasion, individual lava-flows and pyroclastic deposits may be so distinct

TABLE 6.1

Examples of Symbols Used for Igneous Rocks on the Survey Maps of Great Britain

Note that this list is not exhaustive, and that there is considerable variation in the symbols used on individual maps, according to circumstances.

INTRUSIVE ROCKS

G	Granite, Granodiorite	F	Felsite
π	Pegmatite	q^F	Quartz Porphyry
S	Syenite	P	Porphyrite
H	Diorite	D	Dolerite
E	Gabbro	L	Lamprophyre
U	Ultrabasic Rocks	μ	Serpentinite

EXTRUSIVE ROCKS

R	Rhyolite	RZ	Rhyolitic Tuff
T	Trachyte	TZ	Trachytic Tuff
A	Andesite	AZ	Andesitic Tuff
B	Basalt	BZ	Basaltic Tuff
Z	Tuff		
V	Agglomerate (including vent agglomerate)		

These symbols may be prefixed by lower casement letters which are used as abbreviations to further qualify the rock-name (for example, qD is the symbol used for quartz dolerite).

that they can be mapped as separate units. This would be the case, for example, if only a single lava-flow or pyroclastic deposit was erupted, so it became interbedded with sedimentary rocks. It would then be an easy matter to follow the boundaries of this unit in the field. However, the repetitive nature of volcanic activity usually produces lava-flows and pyroclastic deposits which are so similar to one another that they cannot be distinguished as the products of separate eruptions. This means that the various formations mapped in volcanic rocks are mostly formed by a whole series of lava-flows or pyroclastic deposits, which were erupted during a particular phase of igneous activity.

The boundaries between these formations would mark major changes in the nature of the volcanic activity. For example, a thick sequence of andesite lavas might be mapped as one formation, overlain by a series of acid ignimbrites, which represents another formation. Even so, individual lava-flows often do not extend for any distance, while the accumulation of pyroclastic rocks may only occur in the immediate vicinity of a particular vent.

Volcanic rocks are often so heterogeneous that it is impossible to divide the sequence into different formations. The whole sequence of lava-flows and pyroclastic rocks then forms a single formation. This is often divided into a large number of local members, which may represent the products of individual eruptions. It is these members which are defined by the boundaries shown on a geological map.

Contrasts with Sedimentary Formations.

The layered nature of lava flows and pyroclastic deposits obviously means that a lithostratigraphy can be erected for volcanic rocks, using the same methods as applied to sedimentary rocks. Sedimentary bodies tend to die out if they are traced far enough, while there are often breaks in the stratigraphic record, caused by episodes of non-deposition, as described in Chapter 7. Both these features are much exaggerated in the lithostratigraphy developed by volcanic rocks.

Firstly, the very nature of lava-flows and ash-falls means that they tend to form lenticular bodies, which thin to zero as they are traced away from their source. This is particularly the case for viscous lava-flows and coarse ash-falls. Furthermore, such deposits are often erupted from more than one vent during the volcanic history of a particular region. This means that the stratigraphic units formed by these lava-flows and ash-falls often show marked lateral and vertical variations in stratigraphic sequence, particularly if the lava supplied by different vents varies in composition, as commonly found. By way of contrast, fluid lava-flows and ash-flows may be traced for much greater distances from their source, assuming that there was sufficient material available to be erupted over a wide area. The stratigraphic units formed by these deposits tend to be laterally persistent, even although they may show marked variations in vertical sequence.

Secondly, volcanic eruptions typically occur as short episodes of intense activity, separated by much longer intervals of quiescence. Thus, the eruption of lava-flows and pyroclastic deposits tends to be accomplished in almost a single instant of geological time. It is the contacts between these lava-flows and pyroclastic deposits which represent nearly all the geological time required for the accumulation of the volcanic sequence as a whole. This is clearly demonstrated by the long-continued weathering which commonly affects the upper surface of a continental lava-flow before it is covered by the over-lying rocks.

Although the outcrop pattern developed by volcanic rocks can be interpreted in terms of a lithostratigraphy, according to how the various stratigraphic units are arranged in relation to one another, there is usually little or no evidence which would allow a detailed chronostratigraphy to be erectd. Thus, most lava-flows are sufficiently alike that they cannot be distinguished as individual units, able to be traced for long distances in the form of time planes. However, some measure of control is provided wherever different formations interfinger with one another, since each interdigitation would be formed by a one or more lava flows or pyroclastic deposits. Otherwise, it is only the presence of ash-flows in the form of ignimbrite sheets which provides an adequate basis for erecting a detailed chronostratigraphy, since individual sheets can be traced for long distances as time planes, separating stratigraphy units formed during different intervals of geological time from one another.

Igneous Intrusions and their Form

Igneous rocks also occur as intrusions into the pre-existing rocks of the earth's crust. These intrusions vary considerably in shape and size, while they can be formed by a wide variety of rock types. However, they always have one feature in common. The igneous rocks forming such an intrusion are always younger than the country-rocks which they intrude. These country rocks form the walls of the igneous intrusion, so that they are sometimes known as the wallrocks. They may be sedimentary, igneous or metamorphic rocks which bear no relationship to the igneous rocks of the intrusion. Alternatively, an igneous intrusion may itself be surrounded by rocks which were formed during an earlier but related phase of igneous activity. This would be the case, for instance, if an igneous complex was formed by the repeated intrusion of magma into a particular region of the earth's crust. This complex would be formed by a whole series of igneous intrusions, which could be placed in chronological order according to their structural relationships with one another, so allowing an intrusive history to be determined.

Mapping of Igneous Intrusions.

The outcrop of an igneous intrusion is usually shown on a geological map by means of a special ornament or colour, as indicated by the legend or explanation. Symbols can be used to identify not only the rock type forming the intrusion but also its stratigraphic age if this is known. However, the maps published by the Geological Survey of Great Britain only give the rock name in an abbreviated form, starting with a capital letter as shown in Table 6.1. The stratigraphic age is shown on the legend accompanying the geological map. The maps published by the United States Geological Survey use symbols formed by a capital letter, giving the stratigraphic age of the intrusion as shown in Table 2.1, followed by an abbreviated form of the rock name.

The contacts of an igneous intrusion are rarely shown as distinct boundaries on a geological map, unless they are formed by fault lines, since it can be assumed that such a body would necessarily have intrusive contacts with the surrounding country-rocks. Indeed, this may be obvious from the discordant nature of such a body, as shown by the outcrop pattern. However, this assumption does present a difficulty wherever the country-rocks are formed by the igneous rocks of an earlier intrusion. The boundary between these intrusions would only represent the intrusive contact of the later intrusion against the rocks of the earlier intrusion. The two-sided nature of such a contact is only shown occasionally on a geological map by means of a special ornament. Otherwise, it must be inferred from the nature of the outcrop pattern, so allowing these intrusions to be placed in chronological order. It is always the contact of the earlier intrusion with its country rocks, which is truncated by the later intrusion. Alternatively, the outcrops may be considered, since it is the outcrop of the earlier intrusion which is truncated by the later intrusion. For example, a dyke would be earlier than a granite if the outcrop of the dyke was truncated by the contact of the granite with its country rocks, whereas the dyke would be later than the granite if it cut across this contact to intrude the granite itself, as shown in Figure 6.5. It should be noted that it is the later intrusion which preserves its original form, so that the problem is considerably facilitated if this can be determined from the nature of the outcrop pattern. Such conclusions can be supported by field evidence, since it is the later intrusion which would have a chilled margin along its contact with an earlier intrusion, while any pre-existing structures in the earlier intrusion would be truncated along this contact. Thus, wherever a series of igneous intrusions are in contact with one another, they can generally be placed in chronological order. Such intrusions must be younger than the sedimentary rocks which they intrude.

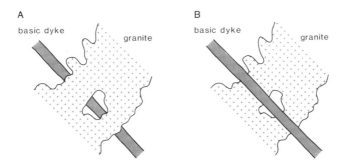

Fig. 6.5. Cross-cutting relationships as shown by igneous intrusions differing in age. A: Basic dyke predates granite. B: Basic dyke postdates granite.

Interpretation of Intrusive Form.

The shape of an igneous intrusion can be considered in relation to its outcrop at the earth's surface, as shown by a geological map. This provides a more-or-less horizontal cross-section through the igneous body at the present level of erosion. The geological boundaries shown on the geological map represent the intrusive contacts developed by the igneous body against the country rocks. How these contacts dip in relation to one another around the circumference of the igneous body can then be used to reconstruct its form in three dimensions. Obviously, it is necessary to determine the attitude of these contacts before such a reconstruction can be attempted.

The attitude of an intrusive contact may be recorded as a structural observation on the geological map, as shown in Figure 6.6. A special dip-and-strike symbol would be used, as indicated in the legend. Such observations can generally be made wherever an intrusive contact is exposed in the field,

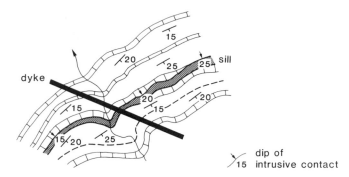

FIG. 6.6 Effect of the topography on the outcrop of an igneous intrusion. Note that the sill forms a concordant intrusion only slightly transgressive to the bedding while the dyke forms a steeply inclined discordant intrusion unaffected by the topography.

although they are rarely shown on geological maps for some reason. If a sufficient number of readings are taken over a wide enough area, it is an easy matter to reconstruct the original form of the igneous intrusion, assuming that these contacts do not show any marked changes in attitude as they are traced above and below the present level of erosion. Naturally, there is a limit to such an extrapolation, beyond which the igneous body might change its shape.

The effect of the topography can also be considered in relation to the geological boundaries defining the outcrop of the igneous intrusion, using the methods already described. This would allow the attitude of an igneous contact to be determined, at least approximately, wherever the geological boundary representing this contact crosses a topographic feature, such as a valley (see Fig. 6.6). This method has the advantage that the form of the igneous body can be determined over a certain interval of height, corresponding to the relief shown by the topography. This allows local irregularities to be recognised in the attitude of an intrusive contact, which cannot be distinguished so easily from the mere recording of structural observations. Moreover, if sufficient information is available over a wide enough area, it may be possible to construct a series of structure contours on the contact of an igneous intrusion, so defining its form in some detail.

Concordant and Discordant Intrusions.

The outcrop of an igneous body can also be considered in relation to the structure of the country rocks, as exemplified by the bedding of sedimentary rocks (see Fig 6.6). Igneous intrusions can be classified as concordant or discordant, according to whether or not their contacts are parallel to the bedding. If an intrusion is concordant, the boundaries defining its outcrop must always be parallel to the geological boundaries formed by the bedding of the country rocks. This means that the igneous body would then have contacts whose attitude could be found from the structural observations giving the dip and strike of the bedding in the country rocks. The intrusive nature of such a body can only be recognised from the outcrop pattern if the structural concordance is not perfect, so that it may be seen locally to cut across the structure of the country rocks.

Taken to the extreme, such features are characteristic of discordant intrusions. Typically, the boundaries representing the contacts of a discordant intrusion cut across the geological boundaries defining the outcrop pattern of the country rocks. However, this need not always be the case, since geological boundaries with the same strike can outcrop parallel to one another, if the relief is sufficiently subdued, even although they may represent cross-cutting contracts in three dimensions, as shown in Figure 6.7.

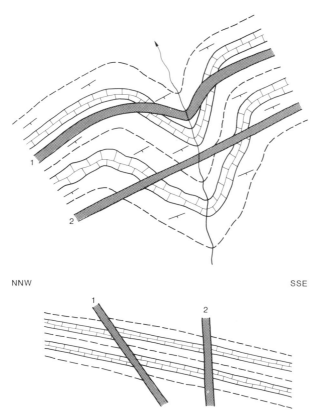

FIG. 6.7. Recognition of discordant intrusions through the effect of the topography on their outcrop. Note that the intrusions appear to be concordant away from the river valley since they have the same strike as the country rocks.

Although a discordant intrusion always cuts across the structure of the country rocks, this may not be clear from the nature of the outcrop pattern. These relationships are important because they provide another frame of reference to reconstruct the form shown by an igneous intrusion, based on the bedding of the country rocks rather than a horizontal plane.

Internal Structures of Igneous Intrusions.

Structural observations may be plotted on the geological map, giving the attitude of various structures developed by the rocks of an igneous intrusion. The nature of these structures is generally shown by the use of a special dip-and-strike symbol, as indicated in the legend or explanation to the geological map. Their attitude may allow the form of the igneous intrusion to be determined, since these structures are commonly developed parallel to its contacts with the country rocks. They are conventionally described as flow structures, since they are assumed to form by lamellar flow as the magma starts to solidify during its intrusion. However, these structures may also develop as a response to the outward pressure exerted by the magma against its wallrocks. They can vary considerably in character.

A flow banding is developed in glassy or fine-grained rocks wherever different layers, varying in composition or texture, are produced by the streaking-out of slight inhomogeneities in the magma as it starts to solidify, as shown in Figure 6.8A. Any tabular or prismatic crystals in the magma may become so aligned that they are now parallel to this flow banding. The parallel alignment of such crystals simply forms flow planes in the absence of any flow banding. Such structures are also developed in viscous lava-flows.

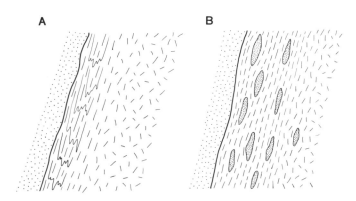

Fig. 6.8. Flow structures in igneous rocks. A: Flow banding in fine-grained or glassy rocks. B: Foliation in coarse-grained rock defined by orientation of tabular or prismatic crystals and the development of flattened xenoliths.

Coarse-grained rocks develop a foliation by the parallel alignment of tabular or prismatic crystals such as mica or amphibole, to form a planar structure in the rock, as shown in Figure 6.8B. Xenoliths of the country-rocks may be flattened in the foliation, forming wispy schlieren as they become incorporated into the magma. Such a foliation can continue to form after the magma has crystallised, provided that the rock remains sufficiently plastic to deform in the solid.

All these flow structures should be clearly distinguished from the compositional layering which can also develop in igneous rocks. This is usually produced by the early formed crystals settling through the magma under the influence of gravity. They accumulate to form layers differing in composition on the floor of the magma chamber. This layering resembles the bedding of sedimentary rocks in that it is horizontal, or nearly so, at the time of its formation. Unless the igneous intrusion forms a horizontal sheet, this layering would not be parallel to its contacts with the country rocks.

Types of Igneous Intrusions

Igneous intrusions vary so widely in shape, size and internal structure that it is not easy to erect a very satisfactory scheme for their classification. A broad division is usually drawn between major and minor intrusions on a basis of their relative size. However, this is rather difficult to define as igneous bodies vary so much in shape, although it might be based on the smallest dimension of the body. It is generally assumed that major intrusions are large bodies with a volume exceeding several cubic kilometres at the very least, while minor intrusions are much smaller bodies. It is possible for major intrusions to be formed by a whole series of minor intrusions, since it is not likely that very large bodies of magma would all be intruded at the same time. This means that major intrusions tend to have an internal structure which is much more complex than that shown by minor intrusions.

The distinction drawn between major and minor intrusions is usually taken to imply that major intrusions are formed by coarse-grained or plutonic rocks, while minor intrusions are composed of relatively fine-grained or hypabyssal rocks. Major intrusions are often known as plutons, in that they are formed by plutonic rocks, although the actual meaning attached to this term is not clearly established in the literature. Major intrusions are usually surrounded by an aureole of thermal metamorphism affecting the country rocks, while minor intrusions often have chilled margins at their contacts. It should be noted that pegmatites occur as extremely coarse-grained rocks in the form of minor intrusions.

Minor Intrusions

The great majority of minor intrusions are tabular bodies with parallel walls, which are generally known as sheet intrusions, intrusive sheets, or sheets. Each intrusion occupies the space which is formed by the opening of a fracture in the country-rocks, as shown in Figure 6.9. Such a fracture is often but not always formed by a pre-existing plane of weakness. The sheet intrusion may follow a particular fracture for a considerable distance, so that it maintains the same attitude throughout most of its course. However, local irregularities may be developed wherever the sheet intrusion encounters cross-cutting planes of weakness, which afford an easier path for the intrusion of magma.

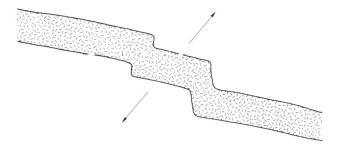

Fɪɢ. 6.9. Intrusive sheet formed by the opening of a fracture in the country rocks.

Sills and Dykes.

A sheet-like body of igneous rock intruded parallel to the bedding of the country rocks is known as a sill (see Fig. 6.10A). There is some doubt if this definition requires the intrusion to be horizontal or nearly so. According to one view, this name should only be applied to a concordant sheet-intrusion if the country rocks are not steeply inclined; otherwise, the intrusion is simply known as a concordant sheet. Although a sill is a typical example of a concordant intrusion, it can show a limited amount of structural discordance by changing its stratigraphic horizon along its length, as shown in Figure 6.10B. It is then termed a transgressive sill. Commonly, this discordance occurs in a series of abrupt steps, between which the sill follows a particular horizon in the normal manner. These steps are commonly formed by pre-existing fault planes. The intrusion of a sill must mean that the magma pressure is sufficient to overcome the weight of the overlying rocks.

A sheet-like body of igneous rock cutting across the bedding of the country rocks, as shown in Figure 6.10C, is known as a dyke (English spelling) or a dike (American spelling). The structural discordance shown by a dyke is essential to its definition, unless it is intruded into massive rocks lacking any internal structure. However, the term is often restricted to steeply inclined or vertical intrusions, implying that the bedding of the country-rocks is horizontal or gently dipping. Dykes rarely occur as single intrusions. Instead, they tend to be associated with one another in large numbers, so forming dyke swarms. These dykes may radiate from an

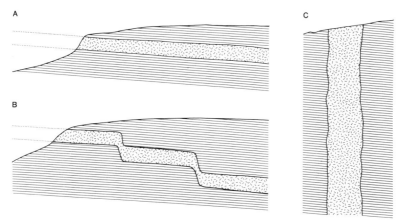

FIG. 6.10. Vertical cross-sections through sheet intrusions. A: Sill; B: transgressive sill; C: dyke.

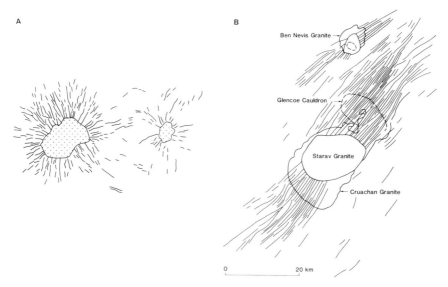

FIG. 6.11. A: Geological map showing the radial dyke-swarms about igneous centres in the Crazy Mountains, Montana (after Weed). B: Geological map showing linear dyke-swarms centred on the Ben Nevis and Etive complexes, Scotland.

igneous centre to form a radial dyke swarm, as shown in Figure 6.11A, or they may all be parallel to one another to form a parallel dyke swarm, which may be focused on an igneous centre, as shown in Figure 6.11B. More complex patterns can also be developed, as shown in Figure 6.12. Dykes are generally formed in response to crustal extension, whether this is caused by the presence of an intrusive complex or whether it is more an expression of regional forces.

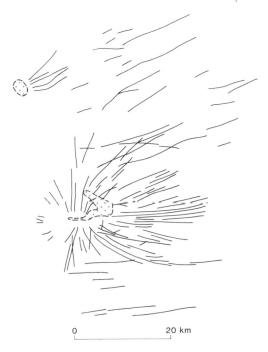

0   20 km

FIG. 6.12. Geological map showing the complex dyke-swarm centred on the Spanish Peaks, Colorado (after Knopf).

Sills and dykes can be distinguished from one another according to the nature of their outcrop, as shown on a geological map(see Fig. 6.13). The definition of a sill as a concordant intrusion means that its outcrop must follow the geological boundaries developed by the country rocks on either side. Accordingly, a sill would be affected by the same structures as the country rocks, as the result of any folding and faulting which occurred after its intrusion. Moreover, if it forms a horizontal or gently dipping sheet, its outcrop would be affected to a considerable extent by the topography. By way of contrast, the definition of a dyke as a discordant intrusion means that its outcrop is likely to cut across the geological boundaries developed by the country rocks on either side. Accordingly, a dyke is rarely affected by the structures developed in the country rocks, unless faulting has occurred after its intrusion. Moreover, if it forms a vertical or steeply dipping sheet, its outcrop would be unaffected by the topography. Only where a dyke has the same trend as the strike of the country rocks would its outcrop in an area of low relief be parallel to the geological boundaries shown by the country rocks (see Fig. 6.7). It would then be difficult to distinguish such a dyke from the outcrop of a sill.

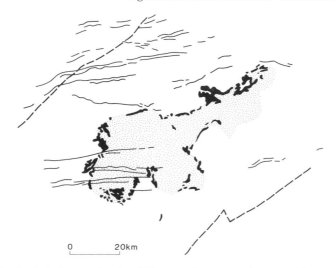

FIG. 6.13. Geological map showing differences in outcrop between dykes and sill-complex. The stippled area represents rocks underlain by the sill-complex.

Sills and Lava-flows.

A sill as a concordant intrusion has a superficial resemblance to a lava-flow, interbedded with sedimentary rocks. However, sills can be distinguished from lava-flows in the field, according to the nature of their upper contacts with the overlying rocks, as shown in Figure 6.14. Lava-flows often have a vesicular or scoriaceous top, which may develop a soil profile as the result of weathering. Irregular masses of sedimentary or pyroclastic rocks may penetrate the lava-flow along cracks in its upper surface, while blocks of the lava-flow may be incorporated as fragments into the overlying rocks. Sills often have a chilled margin of fine-grained and non-vesicular rock along their upper contacts, against which the overlying rocks may be altered as the result of contact metamorphism. These country rocks may also be penetrated by tongues of igneous rock from the intrusion, while they may be incorporated as xenoliths into the underlying sill. Such features would, however, not allow a sill to be distinguished from a lava-flow according to the outcrop pattern, as shown by a geological map. This can only be done if the sill is transgressive, at least locally, so that it changes its stratigraphic horizon as it is traced throughout its outcrop.

Cone Sheets and Ring Dykes.

Sheet intrusions can also occur in the form of cone sheets and ring dykes, centred on an igneous complex. Cone sheets are typically formed by large numbers of arcuate sheet intrusions, which occupy conical fractures inclined at a moderate angle towards a common focus at some depth below

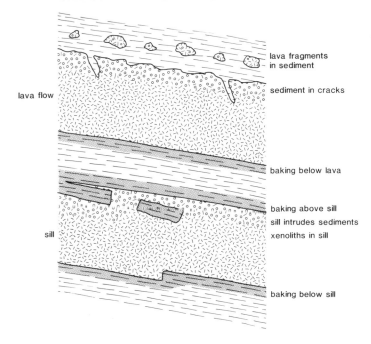

FIG. 6.14. Distictive features shown by lava-flows and sills.

the present surface, as shown in Figure 6.15A. Although they occur in great abundance around the volcanic centres of Tertiary age in Scotland (see Fig. 6.16), they are rarely found elsewhere in the world. Ring dykes are much larger intrusions which occupy steeply inclined fractures around an in-trusive centre, as shown in Figure 6.15B. They tend to have an arcuate out-crop with a diameter of several kilometres, at the very least, so that they belong more to the class of major intrusions. While only a single ring dyke may be present, they mostly occur as a series of concentric intrusions

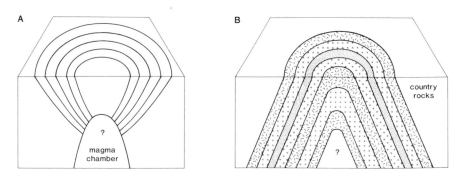

FIG. 6.15. Block diagrams showing the form of (A) cone sheets and (B) ring dykes.

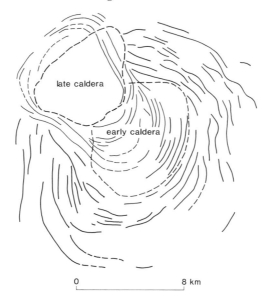

FIG. 6.16. Two sets of cone sheets associated with the Tertiary volcanic centre of Mull, Scotland (after Richey). The sheets have inward dips.

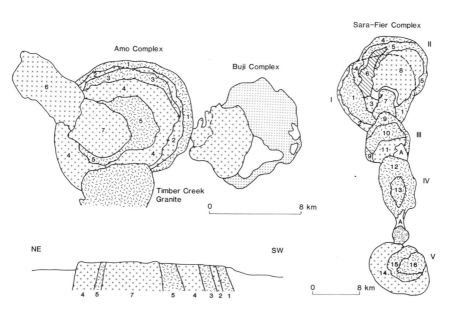

FIG. 6.17. Ring complexes associated with the Younger Granites of Nigeria (after Jacobson, MacLeod and Black, and Turner).

arranged around a common centre, as shown in Figure 6.17. Screens of the country rocks would then be found between the individual ring dykes, unless these intrusions were in contact with one another. Occasionally, a ring dyke is capped by a flat-lying sheet, so that the whole body has the form of what may be termed a bell-jar intrusion.

The space needed for the intrusion of a cone sheet is provided by the uplift of the country rocks lying within its outcrop. By way of contrast, the intrusion of a ring dyke is usually associated with cauldron subsidence, whereby a central block of the country rocks subsides within a ring fracture. The ring dyke is formed by the intrusion of magma into the ring fracture as cauldron subsidence takes place. How easily space can be provided for the intrusion of a ring dyke depends on its inclination, which is often difficult to determine.

If it is inclined outwards, a ring dyke can simply occupy the space formed by cauldron subsidence of the central block. Such a dilational intrusion tends to have a simple ring-like form, since its contacts are formed by the walls of the ring fracture. Obviously, a bell-jar intrusion would be formed if the ring fracture did not reach the surface. This process is known as subterranean cauldron subsidence.

If a ring dyke is vertical or inwardly dipping, a space cannot be provided for its intrusion by cauldron subsidence of the central block. Such a ring dyke may make a space for itself by the piecemeal stoping of the country rocks, which then sink as xenoliths into the underlying magma chamber. However, it seems more likely that a space is provided for its intrusion by an early phase of explosive activity. This would be caused by the escape of volcanic gases, blasting their way upwards from the underlying magma chamber to the earth's surface. The outer contact of such a ring dyke is highly irregular, whereas its inner contact along the ring fracture tends to present a smooth surface against the country rocks lying within the cauldron subsidence. The escape of volcanic gases may be sufficient to convert the magma in the underlying chamber into a liquid-in-gas system. This can then be transported up the ring fracture to erupt as ignimbrite at the surface, so emptying the magma chamber. Once this happens, the central block of country-rocks founders into the underlying magma chamber as the result of cauldron subsidence. This forms a caldera at the earth's surface, closely associated with the eruptions of large volumes of ignimbrite. The final stages in this process is marked by the intrusion of magma around the sides of the cauldron subsidence to form a ring dyke, as the eruption of ignimbrite comes to an end.

Cone sheets and ring dykes can be recognised from the arcuate nature of their outcrop. They may be distinguished from one another since cone sheets have a width of only a few metres whereas ring dykes have a width which often exceeds several hundred metres. Cone sheets are inclined towards the igneous centre at a moderate angle, so that their outcrop may be affected to a certain extent by the topography, whereas ring dykes are steeply inclined, so that their outcrop is unlikely to be affected to any extent by the topography, unless there is considerable relief.

Laccoliths and Domes.

Some minor intrusions do not have a sheet-like form. Laccoliths or laccolites are concordant intrusions which have arched the bedding of the overlying country-rocks to form a low dome with a flat base, as shown in Figure 6.18A. Such an intrusion resembles a sill, except that the magma has not spread laterally for any distance before it solidified. Such a form may reflect the intrusion of highly viscous magma which could not flow sidewards for any distance before it solidified. However, other causes may be responsible for the intrusion of a laccolith rather than a sill. For example, there may be lateral variations in the mechanical or structural properties of the roof rocks, or there may be differences in the weight of the overburden due to topographic relief, which would allow a laccolith to form locally by doming its roof. Commonly, laccoliths have a diameter of a few kilometres or less, while they are rarely more than several hundred metres in thickness. It is usually assumed that the intrusion was supplied with magma from below by means of a pipe-like feeder. However, some laccoliths may be satellite bodies formed by the sideward intrusion of magma from another body, which acts as the feeder.

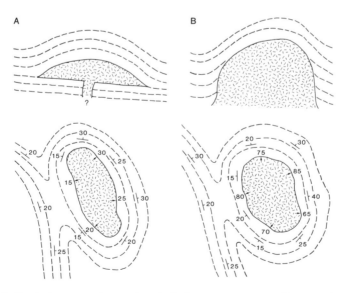

Fig. 6.18. Vertical cross-sections and geological maps showing the difference between (A) laccoliths and (B) intrusive domes.

Laccoliths can definitely be recognised as such in areas of high relief, wherever the topography has cut down to expose the lower contact of the intrusion with the underlying rocks, or where the rocks have been tilted so that the intrusion is seen in cross-section. However, laccoliths are more difficult to recognise in areas of low relief, where the rocks are flat-lying. There, the dome-like form of such an intrusion means that its floor is rarely exposed in cross-section. Accordingly, a geological map would generally only show an outcrop through the roof of the intrusion.

Although this outcrop tends to vary in plan according to the horizontal extent of the intrusion, it generally has a rounded or equidimensional outline. Moreover, the concordant nature of a laccolith requires that the geological boundaries developed within the country rocks around the intrusion should be parallel to its contact. Finally, the bedding of the country rocks would dip away from the igneous body, becoming less steeply inclined as it is traced away from this contact. All these features suggest that the outcrop is formed by a dome-like mass of igneous rock. However, none of these features is sufficient to demonstrate that the upper contact of this intrusion is parallel to the bedding of the country rocks, nor do they show that the intrusion has a flat-lying base, characteristic of laccoliths in general. Similar features could be produced by a discordant dome of igneous rock, against which the bedding of the country rocks is upturned by its forceful intrusion, as shown in Figure 6.18B. Indeed, many intrusions thought to be laccoliths may have such a form.

Volcanic Necks and Plugs.

The pipes or volcanic conduits acting as feeders on a relatively small scale to central eruptions are preserved in the geological record as volcanic necks and plugs (see Fig. 6.19). If such conduits are filled with agglomerate and tuff, they are known as volcanic necks. It is likely that they occupy a space formed by the explosive activity of volcanic gases, escaping from the depths towards the earth's surface. These pyroclastic rocks may be intruded by irregular masses of magma. If a volcanic neck becomes completely filled with igneous rock, it is known as a volcanic plug. Commonly, this forms a single intrusion.

Volcanic necks and plugs are steep-sided bodies, little affected by the topography. They tend to form rounded outcrops, showing a certain amount of irregularity, which represent a horizontal cross-section through the body. Although they are discordant, cutting across the bedding of the country rocks, this feature may not be obvious from the outcrop pattern, as shown on a geological map. Thus, a volcanic plug forming the top of a hill can resemble in its outcrop a sill of igneous rock capping the hill, assuming that the country rocks are flat-lying in both cases, as shown in Figure 6.20. A distinction can be made if the igneous body is repeated elsewhere within

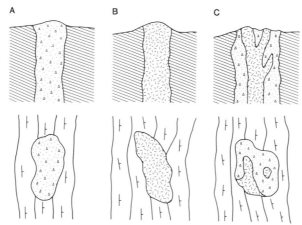

FIG. 6.19. Vertical cross-sections and geological maps showing the form of (A) necks, (B) plugs and (C) composite necks and plugs.

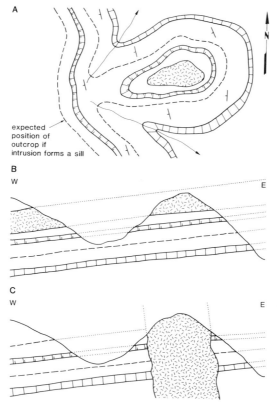

FIG. 6.20. Geologiccal map showing the outcrop of an intrusion which can either be interpreted as (A) a sill or (B) a plug.

the area of the geological map in such a way that it has the form of a sill at the same stratigraphic level; otherwise, the body is more likely to be a volcanic plug.

Such difficulties are not encountered if a volcanic neck is filled with varying amounts of igneous and pyroclastic rocks, since the irregular distribution of these rocks within a well-defined area is sufficiently distinctive for such a body to be recognised from the nature of its outcrop. As the feeder to a volcano, a volcanic neck or plug is likely to pass upwards into a volcanic vent, with less steeply inclined sides. Although volcanic vents are similar in character to necks, the former term should only be used if there is clear evidence that the igneous and pyroclastic rocks lying within its confines were erupted at the earth's surface. This is commonly shown by the development of bedding in the pyroclastic rocks, and the presence of lava-flows within the vent.

Major Intrusions

Whereas minor intrusions can generally be classified according to the various, well-defined categories which have just been described, major intrusions show such a wide variation in shape and size that a simple classification cannot easily be erected. However, a broad distinction can be drawn between the layered sheet-like or funnel-shaped intrusions of basic rocks, which are found in the stable areas of the earth's crust, and the batholithic intrusions of granitic rocks which are found extending to great depths within orogenic belts, forming the eroded roots of fold-mountains.

Lopoliths and Other Intrusions of Basic Rocks.

Major intrusions of basic rock often adopt the characteristic form of a lopolith. This is a sheet-like body of igneous rock occupying a structural basin, so that it has the form of a saucer, as shown in Figure 6.21. It is generally concordant with the structure of the country rocks. Lopoliths are commonly several kilometres in thickness, while they may underlie an area exceeding many thousands of square kilometres in extent. Such a large body of igneous rock often has a very complex history, which is difficult to decipher. However, lopoliths are generally formed by layered sequences of basic and ultrabasic rocks, passing upwards into acid rocks. Such a complex may be the result of more than one intrusive episode, so that the lopolith is a multiple intrusion. It can be overlain by lava-flows and pyroclastic deposits, which are related in some way to its intrusion.

Although defined as a saucer-shaped body, there is some evidence that lopoliths were fed from below by narrow, funnel-shaped intrusions, into which they pass as they are traced to depth. These funnel-shaped intrusions may have a dyke-like form with a width of a few kilometres, becoming

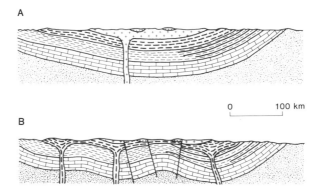

Fig. 6.21. Vertical cross-sections showing the inferred structure of a lopolith. A: Original interpretation. B: Recent revision of the original interpretation.

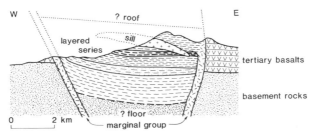

Fig. 6.22. Vertical cross-section showing the funnel-shaped form of the Skaergaard intrusion, Greenland (after Wager and Brown).

narrower at depth. It should be noted that lopoliths and their feeders are only represented by a few examples in the world, so making any generalisations difficult to establish in view of their exceptional character.

Basic rocks can also occur as major intrusions with the sheet-like form of sills, rather than lopoliths. Such intrusions may be layered. They are also found as major intrusions in central complexes, representing the eroded roots of basaltic volcanoes. Such intrusions commonly have the form of ring dykes, plugs and stocks, although they may also occur as funnel-shaped bodies, opening upwards, as shown in Figure 6.22. The commoner forms adopted by such intrusions have already been described under the heading of minor intrusions.

Classification of Granitic Intrusions.

The wide variety of form adopted by granitic intrusions means that a simple classification of such bodies is difficult to erect. It is commonly the case that granitic intrusions on a minor scale occur in the form of dykes, sills, sheets, laccoliths, plugs and more irregular bodies, which generally come into the class of minor intrusions. However, the larger intrusions of granitic rocks appear to form igneous bodies which, extending to great depths in the earth's crust, appear to lack any foundation. In particular, the contacts of such intrusions appear to dip outwards so that the igneous body would become wider at depth without any evidence of a floor to the intrusion, as shown in Figure 6.23. Such a body is known generally as a batholith if it outcrops over an area exceeding 100 square kilometres (about 40 square miles). An intrusion of a similar form to a batholith, outcropping over a smaller area than 100 square kilometres, is termed a stock.

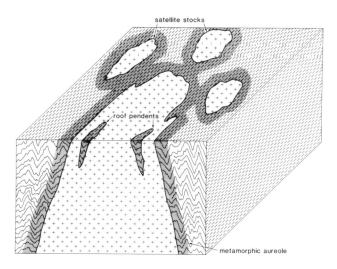

Fɪɢ. 6.23. Block diagram showing the inferred form of a granite batholith.

It should be emphasised that batholiths are often very much larger than this limiting figure, extending for many hundreds of kilometres along strike with a commensurate width. There is a tendency to restrict the term "batholith" so that it would only apply to such a large body of granitic rocks, which is often formed by a whole complex of separate intrusions on a smaller scale. Batholiths and stocks tend to outcrop as rounded or elongate masses, parallel to the structural trends in the country rocks. A stock with a rounded outcrop is known as a boss. The boundaries of these intrusions are often irregular, with tongues or apophyses of igneous rocks penetrating the country rocks.

Isolated outcrops of igneous rock may occur around the margins of stocks and batholiths. If these satellite masses have an underground connection with the main body of the intrusion, they are known as cupolas. The country rocks may also occur as isolated outcrops, surrounded by the igneous rocks of the intrusion. Such masses of country rock are known as roof pendants if they are formed by the roof of the stock or batholith, projecting downwards into the igneous body. However, they might be formed by detached masses of country rock, which have foundered as large-scale xenoliths into the intrusion, so that they would be surrounded on all sides by igneous rock. A distinction between roof pendants and large-scale xenoliths can only be made if these detached masses became structurally disorientated as they sank into the magma, so destroying the stratigraphic continuity between these masses and the country rocks surrounding the intrusion. Otherwise, the bedding of the country rocks would continue through these masses without any change in attitude, whatever their origin.

Permissive and Forceful Intrusions.

Another basis for the classification of granitic intrusions concerns their relationship with the surrounding country rocks. This may also be reflected in the form of their internal structure. A broad distinction can be drawn between permissive intrusions, which occupy space formed by the passive displacement of the country rocks, and forceful intrusions which make space for themselves by shouldering aside the country rocks. The different mechanisms involved in the emplacement of such intrusions are shown in Figure 6.24. The passive displacement of the country rocks is accomplished

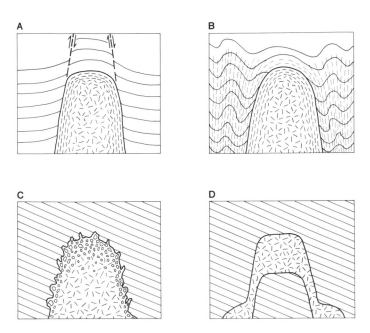

FIG. 6.24. Possible mechanisms involved in the emplacement of granitic bodies. A and B: forceful intrusion; C: piecemeal stoping; D: cauldron subsidence.

by a process known as stoping, whereby masses of the pre-existing rocks are detached along fractures from the walls and roof of the intrusion, so forming a space for the magma to occupy. This process is known as piecemeal stoping if it occurs on a small scale. It produces xenoliths of country rocks, which can then sink to depth within the magma chamber. A similar process operates on a larger scale to produce ring dykes through the mechanism of subterranean cauldron subsidence.

Ring Complexes and Cauldron-subsidence.

Some granites are intruded in the form of ring complexes at a relatively high level in the earth's crust. They generally occur as circumscribed masses with annular outlines, formed as a result of cauldron subsidence. It has already been described how the intrusion of a ring dyke may be related to the formation of a caldera at the earth's surface. Such a ring dyke commonly forms an annular intrusion of granitic rock, dipping inwards at a steep angle around the margin of a cauldron subsidence. The sunken block of country rocks would be capped in many cases by the lava-flows and pyroclastic deposits which accumulated within the caldera.

Although cauldron subsidence may occur more than once, forming a series of ring dykes, the central block of country rock would be retained as a characteristic feature of such a complex. However, such a central block of country rocks is lacking in most ring complexes formed by granitic rocks. Instead, each complex is formed by a concentric series of annular intrusions, which generally become younger as they are traced towards its centre. If such a complex is formed by cauldron subsidence, the ring intrusions must dip outwards, so allowing a space to be formed by the subsidence of a central block. In general, such intrusions are most likely to be formed as the result of underground cauldron subsidence, so that they would have the form of bell-jar intrusions, with steep sides but flat tops.

It must be admitted that there are two difficulties to be faced by such a hypothesis. Firstly, the central block needs to subside by a very considerable amount to provide sufficient space for the intrusion of a ring complex as a whole, unless the ring intrusions dip outwards at a much shallower angle than often appears to be the case. Secondly, although the sunken block of country rocks should be preserved at depth within the ring complex, such a block is rarely if ever exposed. This may suggest that the outermost intrusions, which was the first to form, was not a ring dyke at all but a stock of some sort, so allowing the cauldron subsidence to be formed by a sunken block of igneous rocks. It should also be noted that the ring-like form of the various intrusions in a ring complex is not necessarily an original feature, since each intrusion is generally in contact with younger rocks along its inner margin. Only if the country rocks are preserved as screens between the different intrusions can it be demonstrated that they were originally intruded as ring dykes.

Intrusions formed by Piecemeal Stoping.

High-level granites can also occur as discordant intrusions of igneous rock, lacking any internal structures and cutting across the structure of the country rocks. Typically, they occur as rounded or elongate masses, which are often somewhat irregular in outline, as shown in Figure 6.25. Roof pendants and satellite stocks may be common, suggesting that the roof to such an intrusion lies close to the present level of erosion. Although the contacts of these bodies against the country rocks tend to be irregular, they usually appear to be inclined outwards at a steep angle, so that the intrusion itself would become wider and more extensive as it is traced to greater depths. There is usually no evidence that a foundation of older rocks forms the floor to such an intrusion. How it terminates at depth is, therefore, a matter of inference.

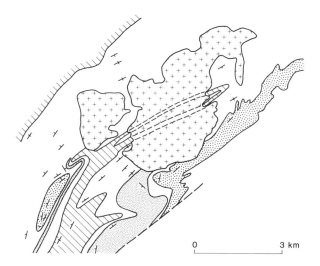

Fɪɢ. 6.25. Geological map showing the form of Mullach nan Coirean Granite, Scotland.

It is commonly found that the regional structures developed within the country rocks show no signs of disturbance as they are traced towards the contact of such an igneous body. It is generally supposed that a space has been made for the intrusion by piecemeal stoping, whereby blocks of the country rocks were detached from the walls and roof of the intrusion to form xenoliths, which sank into the magma as it invaded the country rocks. A difficulty facing this hypothesis is that many bodies of granitic rock, which apparently can only have been formed by piecemeal stoping, lack appreciable numbers of xenoliths, while there is often little evidence that these xenoliths have been assimilated into the magma before it crystallised.

It has therefore been suggested that these xenoliths sank to such depths that they have simply disappeared from view.

However, it would still be expected that some intrusions would be exposed, representing the roots of these bodies, composed of igneous rocks full of xenoliths or showing evidence of their assimilation. The apparent absence of such bodies in the geological record, unless they are represented by xenolithic intrusions of dioritic rock, means that there is a space problem concerning such discordant intrusions formed by granitic rocks. It is possible that these rocks were formed in the solid state by the granitisation of pre-existing country rocks. The country rocks would then be converted or transformed into granite by the influx of various elements as the result of solid diffusion, coupled with the loss of other elements. However, the geological and geochemical evidence now does not appear to support such a hypothesis.

Ring complexes and discordant intrusions share a number of features in common. They are usually formed by massive rocks lacking any foliation. They commonly have sharp contacts, cutting across the structures in the country rocks. Although these country rocks are often folded to a considerable extent, they are not usually affected by a high grade of regional metamorphism. Indeed, many ring complexes and discordant intrusions are intruded into non-metamorphic rocks. Such intrusions commonly develop a wide zone of thermal metamorphism affecting the country rocks, giving rise to a metamorphic aureole around their outcrops. However, such bodies may simply develop a chilled contact against the country rocks, which would then be unaffected by any degree of thermal metamorphism.

Forceful or Diapiric Intrusions.

Granitic intrusions can also occur as forceful intrusions, which have made a space for themselves by pushing aside the country rocks as they were intruded. Such an intrusion usually outcrops as a round or elongate mass with smooth outlines, as shown in Figure 6.26, developing a lobate form wherever the contact is somewhat irregular. Roof pendants are nearly always lacking, although satellite stocks may be present. The forceful nature of such an intrusion is shown by the way that the country rocks become distorted as they are traced towards the igneous body. While this occurs in such a manner that the regional structures tend to become parallel to the contact of the igneous intrusion, it also means that new structures are developed in the surrounding country rocks. These structures are commonly represented by folds and their axial-planar cleavages. If the country rocks were originally flat-lying, the folding may result in the arcuate development of a rim syncline, which can often be traced around the perimeter of such an intrusion. Local structures of this sort would only modify the regional

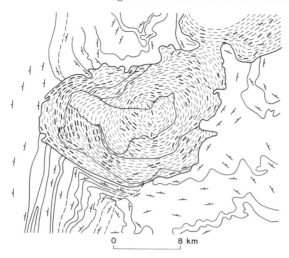

Fɪɢ. 6.26. Geological map showing the outcrop of the White Creek Granite, British Columbia (after Buddington).

structures developed elsewhere in the country rocks, away from the intrusion. However, where such intrusions occur close together, it is commonly found that the form and arrangement of these igneous bodies controls the nature of the regional structures developed within the country rocks.

Although the central parts of a forceful intrusion may be composed of massive rock, a foliation is commonly developed towards its margins with the country rocks. In general, this foliation is formed parallel to the contacts of the igneous body. Its development is closely related to the formation of axial-planar cleavages in the country rocks, associated with the folding and deformation that occurs in response to the forceful intrusion of the igneous body. Such structures can only develop if the country rocks are capable of deforming plastically under the influence of temperatures and pressures appropriate to regional metamorphism. Since the country rocks would then be affected by regional metamorphism, a distinct aureole of thermal metamorphism is rarely developed around a forceful intrusion. However, the intensity of regional metamorphism may increase as the country rocks are traced towards its contact. Such a contact often appears to be gradational in character, since the juxtaposition of foliated granite against high-grade metamorphic rocks commonly occurs across a zone where sheets, veins and irregular masses of granitic rock are present. Such a mixture of igneous and metamorphic rock is known as a migmatite.

Distinctive Features of Granitic Intrusions.

Ring complexes, discordant intrusions and forceful intrusions can be distinguished from one another according to the nature of the outcrop pattern, as shown by a geological map. Ring complexes can be recognised since they outcrop as a concentric series of annular or arcuate intrusions, cutting across the structures of the country rocks. Whether the contacts of these ring intrusions dip inwards or outwards can usually only be determined if the outcrop of the geological boundaries is affected by topography. Discordant and forceful intrusions resemble one another in that they tend to outcrop as rounded or elongate masses. However, discordant intrusions may be recognised since they cut across the regional structures of the country rocks in such a way that these structures may be traced towards the igneous body without any disturbance, as shown in Figure 6.25. Forceful intrusions can be distinguished from discordant intrusions since the regional structures become distorted as they are traced towards the igneous body (see Fig. 6.26). This occurs so that the strike of the country rocks becomes more-or-less parallel to the igneous body at its contact. This means in effect that forceful intrusions can generally be classified as concordant bodies to a certain extent. However, local discordances are usually present, thus demonstrating the intrusive nature of such bodies. It should be emphasised that there appears to be a complete gradation between discordant and concordant plutons, depending on the relative importance of piecemeal stoping *vis-à-vis* forceful intrusion as the mechanism of emplacement.

The three-dimensional form of a discordant intrusion can generally be determined only if the topography has sufficient relief to affect the outcrop of its contact with the country rocks. However, such evidence would be supplemented in the case of a forceful intrusion by structural observations giving the attitude of the foliation developed within the igneous body parallel to its contact, and by structural observations giving the attitude of bedding and other structures in the country rocks, which would also be more-or-less parallel to this contact. It is generally found that such intrusions form bulbous masses which become wider at depth. However, some examples of forceful intrusions appear to become narrower at depth, suggesting that they have become detached from their roots as they moved upwards into their present position. Finally, some intrusions of granitic rocks have a sheeted structure, formed by the wedging apart of the country rocks by dyke-like masses which then coalesced to form what now appears to be a single intrusion.

CHAPTER 7

Unconformities and the Geological Record

Breaks in Stratigraphic Sequence

THIS chapter deals with angular unconformities as a particular type of stratigraphic break which interrupts the deposition of sedimentary rocks. Such a break in the stratigraphic record occurs as a result of earth movements. However, it is clearly related to other types of stratigraphic break which simply result from the non-deposition of sedimentary rocks, combined in some cases with erosion of the underlying rocks, as shown in Figure 7.1. These breaks in the stratigraphic record are known variously as diastems, non-sequences, paraconformities and disconformities. They represent intervals of geological time when sedimentary rock was not deposited in a particular area.

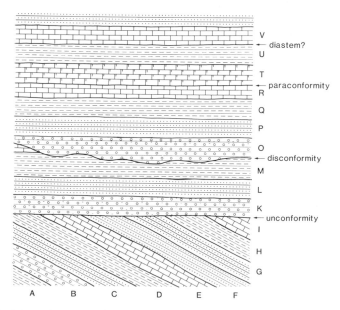

FIG. 7.1. Schematic cross-section showing the various types of break in the stratigraphic record. The letters refer to an orderly sequence of biostratigraphic zones.

Paraconformities and Disconformities.

It is unlikely that sedimentary rocks were ever deposited in a uniform manner over even a short interval of geological time. Indeed, the development of bedding as a characteristic feature of sedimentary rocks indicates that their deposition was marked by changes in the nature of the sediment being deposited, fluctuations in the supply of this sediment, changes in the conditions of deposition, pauses in sedimentation and so on. Some bedding planes are probably formed in response to abrupt changes in sediment supply and the environmental conditions. However, other bedding planes are more likely to represent longer pauses in sedimentation. Such bedding planes are termed diastems if they are not marked by any feature which would allow them to be recognised. It is quite possible that these bedding planes could represent a longer period of geological time than that required for the deposition of the intervening beds. This would be the case if the time taken to deposit the individual beds of a stratigraphic sequence was much less than the time represented by the sequence as a whole.

Longer periods of non-deposition can be recognised in the stratigraphic record if they result in the absence of rocks belonging to one or more biostratigraphic zones. This would be marked by the lack of diagnostic zone fossils at a particular horizon. Such periods of non-deposition may also be recognised by the effects of weathering on the underlying rocks, by the evidence of organic activity in the form of animal burrows, tracks and trails, by the development of root zones in the underlying rocks, by the accumulation of phosphate and manganese nodules, and by the preservation of derived fossils belonging to more than one biostratographic zone. Breaks of this magnitude are termed non-sequences or paraconformities if they are simply the result of non-deposition. However, if the break is marked by the eriosion of the underlying rocks, the erosion surface forming the base of the overlying rocks is termed a disconformity, provided that the beds above and below the erosion surface still remain conformable to one another on a large scale. The erosion surface may be highly irregular on a small scale. Paraconformities and disconformities differ from diastems in that they can be recognised.

Definition of an Unconformity.

In contrast to a disconformity, an unconformity is produced wherever the rocks lying below a surface of erosion have been tilted or folded as the result of earth movements, prior to the deposition of the overlying rocks. These earth movements may be accompanied by faulting, igneous intrusion or regional metamorphism. Since it can be assumed that the underlying rocks were deposited as horizontal beds, a structural discordance would exist between the bedding of these rocks and the erosion surface forming the base of the overlying sequence. An unconformity is then said to occur between the two sequences, such that the overlying beds rest unconformably on the underlying rocks. This term is also applied to the erosion surface separating the two sequences from one another, unless it is distinguished as the surface of unconformity. The meaning attached to the term "unconformity" can usually be determined from the context in which it is used to describe either an uncomformable relationship between two sequences or the erosion surface separating these sequences from one another. Such an erosion surface forms an unconformable contact between the two stratigraphic sequences, in contrast to the conformable contacts which are otherwise developed between sedimentary formations.

This definition identifies an unconformity as a structural break. However, the recognition of an unconformity implies a certain history of geological events, whcih can be interpreted by assuming that sedimentary beds are usually deposited with little or no dip. Unless the surface of erosion represents a buried landscape with a certain amount of topographic relief, it is usually found that the rocks above an unconformity were deposited with the bedding more-or-less parallel to the unconformity. This implies that the unconformity was originally a horizontal surface of erosion, cutting across the structures developed in the underlying rocks. Since erosion can only occur above the base level of erosion, while sedimentary rocks tend to accumulate below this level, uplift must have occurred after the underlying rocks were deposited, so allowing erosion to take place. It is usual to assume that this uplift occurred after the folding of the underlying rocks, while it must have ceased before the erosion of the underlying rocks came to an end. Thus, the presence of an unconformity marks an interval of geological time when the underlying rocks underwent folding, uplift and erosion, prior to the deposition of the overlying rocks, as shown in Figure 7.2. It may be noted that the renewed deposition of sedimentary rocks on top of an unconformity generally occurs once subsidence takes the place of uplift during the geological histoy of a particular region.

Differences in Terminology.

It was originally in the sense of a structural break that the term "unconformity" was introduced into the British literature, derived from the latin noun "forma" meaning shape or form. This usage is still followed in Great Britain. However, the meaning of this term has been extended in North America so that it is applied to any gap in the stratigraphic record, marked by the non-deposition of sedimentary rocks over an appreciable length of geological time. Parconformities and disconformities are, therefore, included under the heading of unconformities, even although the overlying beds are deposited conformably on top of the underlying rocks in such

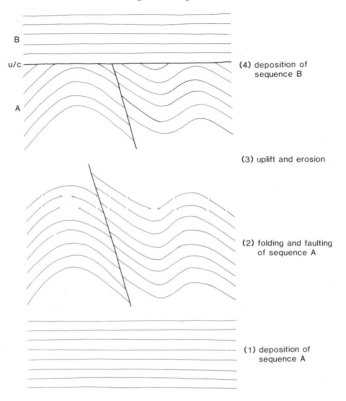

FIG. 7.2. Stages in the development of an angular unconformity (u/c) between two stratigraphic sequences A and B.

cases. If this definition of an unconformity as a temporal break is adopted, the type of unconformity marked by a structural discordance between sedimentary sequences must be distinguished as an angular unconformity. It may be noted that an unconformity formed by sedimentary rocks resting on top of massive igneous or metamorphic rocks is sometimes known rather pedantically as a non-conformity.

Table 7.1 summarises the salient features of the various breaks in the stratigraphic sequence according to the British terminology. It is possible

TABLE 7.1

British terminology	Non-deposition	Uplift and erosion	Deformation
Diastem	Yes	No	No
Paraconformity	Yes	No	No
Disconformity	Yes	Yes	No
Unconformity	Yes	Yes	Yes

for such breaks to grade laterally into one another. For example, a non-sequence changes into a paraconformity if such a break in the stratigraphic sequence is marked by a longer interval of non-deposition, a paraconformity changes into a disconformity if non-deposition is associated with uplift and erosion, and a disconformity changes into a unconformity if uplift and erosion are accompanied by tilting or folding of the underlying rocks.

Overstep

How an unconformity cuts across the structures developed in the underlying rocks is described in the British literature as overstep. It is not a term much used in North America, where "overlap" has been used in the same sense. However, this term has an entirely different meaning in the British literature, to be described later in this chapter. Overstep will therefore be used in the British sense. It describes the way in which the rocks lying above an angular unconformity truncate the upturned and eroded edges of the underlying beds. The overlying rocks are said to overstep the underlying beds along the plane of unconformity, while it is the rocks below the unconformity which are overstepped in this manner. Overstep defines the structural relationship which exists between an unconformity and the underlying rocks.

Many unconformities simply represent horizontal surfaces of erosion, cutting across the structures developed in the underlying rocks. It can be assumed as usual that the sedimentary rocks lying below an unconformity were originally deposited as horizontal beds. The structural discordance now existing across an unconformity must therefore be a direct measure of the earth movements which have affected the underlying rocks, prior to the deposition of the overlying beds. The overstep shown by such an unconformity can therefore be used to determine the nature of these pre-existing structures.

Tilting and Its Effects.

The simplest type of angular unconformity is formed wherever the underlying rocks are affected by tilting movements, prior to the deposition of the overlying rocks. Such movements commonly affect fault blocks of basement rocks, extending widely over a whole region, beyond which a series of marginal flexures may be developed. Only a very slight degree of structural discordance is then required for an angular unconformity to be developed on a regional scale. It is usually the case that such an unconformity cannot be recognised in a single exposure, since the difference in dip between the two sequences may well be imperceptible. However, traced over a wide enough area, it is generally found that the sedimentary rocks lying above such an unconformity gradually overstep different formations within the underlying sequence, so allowing the unconformity to be recognised, as shown in Figure 7.3A. This type of overstep can be observed on a regional scale wherever the sedimentary formations are relatively thin bodies, extending over a wide area.

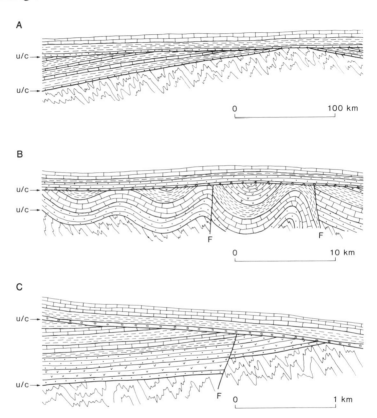

Fig. 7.3. Stratigraphic cross-sections showing the overstep associated with different types of unconformity, according to the deformation affecting the underlying rocks. A: tilting; B: folding; C: faulting. Each cross-section shows the lower stratigraphic sequence to be underlain unconformably by basement rocks. Note differences in horizontal scale between each cross-section.

For example, a sedimentary formation with a thickness of 100 m would be overstepped within a distance of 10 km, even if the difference in dip between the two sequences was slightly less than 0.5°. Acccordingly, a thickness of 1000 m would be removed by erosion from the underlying sequence if such a slight difference in dip was maintained over a distance of approximately 100 km. Such a relationship suggests that sedimentary rocks are deposited with an initial dip which is virtually zero, since it implies that the depositional surface needs only to be distorted by a very small amount for erosion to take the place of deposition.

Effect of Folding and Faulting.

If the overstep is simply the result of folding, older rocks will be overstepped by the unconformity wherever an anticline is developed in the underlying rocks, whereas younger rocks will be encountered in the opposite direction, towards the core of a complementary syncline, as shown by Figure 7.3B. The form originally displayed by these folds can be reconstructed from the nature of the angular discordance which now occurs across the unconformity, whether or not the overlying rocks have subsequently been folded or faulted. This discordance increases from zero to a maximum, as the unconformity is traced away from the crest or trough of a fold developed in the underlying rocks. Accordingly, the angular discordance associated with an unconformity is most marked wherever upright and rather tight folds are developed in the underlying rocks. Such an unconformity can easily be recognised from observations in a single exposure. The angular discordance would be less marked if the folds developed in the underlying rocks were inclined or recumbent structures, provided that they were tight or isoclinal, or if they were upright folds of a more open nature.

Overstep has a progressive character if it is simply the result of tilting or folding. The various formations would then be overstepped in stratigraphic order, or its reverse, as the unconformity cuts across the structures developed in the underlying rocks. However, this orderliness tends to break down wherever faulting has occurred, prior to the deposition of the overlying rocks. The unconformity would then overstep the underlying rocks in such a way that it cuts across the faults developed in response to these movements. If the stratigraphic separation across these faults is sufficiently great, one or more sedimentary formations may be cut out by the faulting. This means that the unconformity would overstep abruptly from one formation to another, which might not be in stratigraphic order, as it cuts across the fault planes developed in the underlying rocks, as shown in Figure 7.3C.

Recognition of Overstep.

The outcrop pattern shown by a geological map allows the presence of an unconformity to be recognised wherever it can be shown that the base of one stratigraphic sequence oversteps a series of sedimentary formations within an underlying sequence. This means that an unconformity can only be recognised where overstep can clearly be demonstrated.

By definition, the sedimentary formations forming a stratigraphic sequence on top of an unconformity are conformable with one another. They outcrop in a regular manner, only disturbed by the folding and faulting that may have affected this sequence after it was laid down. In particular, the outcrop of each formation within this sequence is always flanked on either side by two other formations, so that the same sequence of stratigraphic formations is preserved throughout, unless there are any lateral changes in sedimentary facies. Once such a sequence has been recognised, it is possible to determine its stratigraphic order, using the methods already described.

In general, the bedding within a sequence of stratigraphic formations dips away from its base, unless it has been overturned. The oldest formation within this sequence can be recognised since, although it is part of this stratigraphic sequence, its lower contact cuts across the geological boundaries developed between the underlying formations. This contact then represents the unconformity at the base of the overlying sequence. Such an unconformity has a cross-cutting relationship with the underlying rocks simply because there is a structural discordance between the two sequences.

This means that the boundaries developed by the overlying rocks are conformable with the line of the unconformity, whereas the boundaries developed within the underlying rocks are truncated by the unconformity itself, as shown in Figure 7.4. Such an outcrop pattern is merely the distorted form of an unconformity, as seen in cross-section. It is typically developed wherever the sedimentary rocks lying above an unconformity dip uniformly away from the unconformity. This would be the case if these rocks had simply been tilted from a horizontal attitude after they had been deposited.

Unconformities become more difficult to recognise if the outcrop of the overlying rocks is affected to a considerable extent by the topography, so forming a series of outliers with irregular boundaries; if the overlying rocks are folded and faulted as well as the underlying rocks, so obscuring the structural differences between the two sequences; or if the structural discordance across the unconformity is only slight, so that the dip and strike of the two sequences do not differ by very much. Under these circumstances, the outcrop pattern needs to be examined very carefully in order to determine whether or not an unconformity is present.

Even so, such an unconformity can only be recognised where it can be shown that the lowermost formation within one sequence comes to rest on top of different formations with an underlying sequence, as is traced throughout a particular region. An example is shown in Figure 7.5. It is this evidence for overstep which is critical to the identification of an unconformity. However, even if overstep cannot be demonstrated on a regional scale, it may be seen locally wherever the overlying rocks have a uniform dip and strike. For example, the unconformity may cut across the bedding of the underlying rocks, as defined by the outcrop of a particular bed such as a coal seam. Alternatively, it may cut across fault lines or igneous intrusions

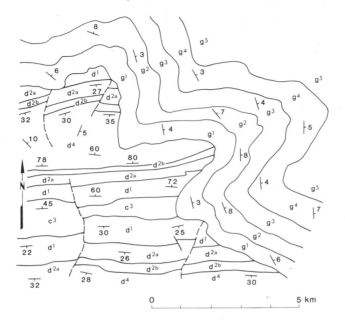

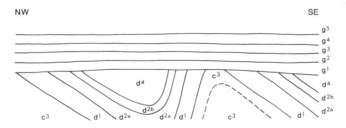

FIG. 7.4. Geological sketch-map and vertical cross-section showing the outcrop pattern characteristic of an unconformity cutting across the formations developed in the underlying rocks.

within the underlying rocks, which do not affect the rocks lying above the unconformity. Local evidence of this sort may be sufficient to demonstrate overstep of the underlying sequence, thus allowing an unconformity to be identified at the base of the overlying sequence.

Overlap

How the overlying rocks are deposited as individual beds on top of an un-conformity is described in terms of overlap, particularly in the British literature. It has the same sense as onlap, which is a term more used in the American literature. Some unconformities show little or no overlap, in that

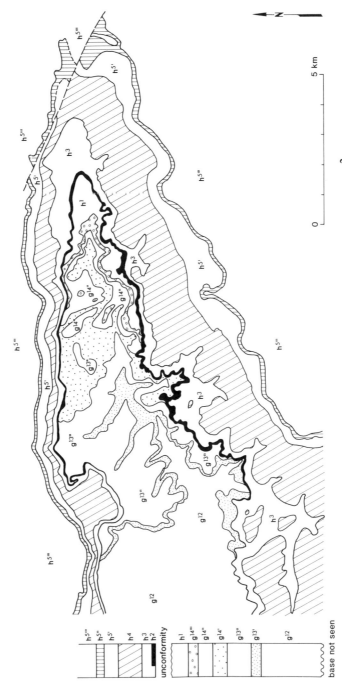

FIG. 7.5. Geological map showing an unconformity developed at the base of the h² formation. (based on Sheets 296 and 297 of the Geological Survey of England and Wales.)

the plane of the unconformity is parallel to the bedding of the overlying rocks. However, other unconformities show a marked degree of overlap, since the surface of unconformity is inclined at an angle to the bedding of the overlying rocks, as shown in Figure 7.6. Each bed would then disappear as its thickness decreases to zero as it comes into contact with the underlying rocks along the unconformity, so forming what is known as a feather edge. This would allow the next bed in the stratigraphic sequence to extend over a wider area, before wedging-out against the unconformity. Accordingly, the higher beds in the stratigraphic sequence are said to overlap the lower beds within the same sequence, as each bed in turn wedges out against the unconformity. It is the lower beds which are overlapped in this way. Overlap defines the stratigraphic relationship between the unconformity and the overlying rocks.

A clear distinction must be drawn between overstep and overlap. This may be borne in mind by remembering that overlap only affects the rocks above an unconformity while overstep occurs between the unconformity and the underlying rocks. Unfortunately, this distinction has not been made until recently in North America, where overlap has been used to describe overstep in the British sense, as well as overlap in the strict sense. The position is further confused since overlap has been termed onlap, as the opposite to offlap.

The interpretation to be placed on overlap depends on the usual assumption that sedimentary rocks are mostly deposited with little or no dip. Thus, the effect of any folding which has occurred after their deposition can be removed by restoring the bedding of sedimentary rocks to the horizontal. It would then be found that the erosion surface representing an unconformity at the base of a sedimentary sequence was inclined at an angle to the horizontal, depending on the degree of overlap shown by the overlying rocks. Accordingly, the overlap shown by the sedimentary rocks lying above an unconformity depends on the nature of the erosion surface represented by the unconformity.

Overlap on a Regional Scale.

If this erosion surface marks a continental peneplain or a flat-lying surface of marine erosion, it would be nearly if not quite horizontal. Since sedimentary rocks would also be deposited with little or no dip on top of this horizontal erosion-surface, the bedding of these rocks would be parallel to the unconformity at their base. Under these circumstances there would be little or no overlap shown locally by the sedimentary rocks deposited on top of the unconformity. However, overlap can still be developed on a regional scale in such a case. This would occur if the erosion surface was even slightly inclined from the horizontal, so allowing deposition to start earlier

FIG. 7.6. Stratigraphic cross-section showing the overlap of various formations within a sedimentary sequence lying above a plane of unconformity.

at one place than another. Deposition would then spread over a wider and wider area as the erosion surface gradually became buried under sedimentary rocks, in such a way that the base of the sedimentary sequence is diachronous.

Overlap of this type is marked by the wedging-out of individual chronostratigraphic zones against the unconformity, so allowing younger and younger rocks to come into contact with the surface of unconformity as it is traced throughout a particular region. This is often associated with the development of a marine transgression across the land, which allows the deposition of marine sediments to occur over a wider area as the sea advances and the shore line retreats. The degree of overlap associated with such a marine transgression depends on how slowly the sea spreads over the land. Overlap would only be marked if the transgression took such a long time that the base of the sedimentary sequence was markedly diachronous.

While overlap is always defined by the wedging-out of chronostratigraphic zones, it may or may not be reflected by the lithostratigraphy. This depends on how the sedimentary facies migrate towards the land as the transgression takes place, since it is these facies which define the various formations in a particular lithostratigraphy. Figure 7.7 shows the two possibilities. Firstly, the sedimentary facies may migrate at the same rate as the marine transgression, so that they keep pace with the advance of the sea over the land. The boundaries of the various formations, as defined by the different types of sedimentary facies, would then be parallel to the underlying unconformity, as shown in Figure 7.7A. The lithostratigraphy would not show any overlap in such a case, simply because the individual formations do not wedge-out against the unconformity.

However, if a gradual change in the distribution of the sedimentary facies occurred as the marine transgression took place, so that there was a change in the nature of the sediment being deposited along the shore line as it migrates towards the land, overlap would be developed by the wedging-out of individual formations against the unconformity, as shown in Figure 7.7B. For example, if shales came to be deposited where sandstones had previously been laid down, the shale formation would overlap the sandstone formation. Likewise, if limestones came to be deposited where shales had

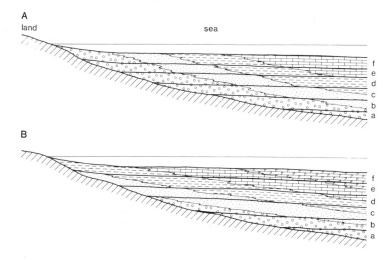

FIG. 7.7. Stratigraphic cross-sections showing how overlap may be defined. A: Overlap of chronostratigraphic zones (a, b, c ...) not reflected in the lithostratigraphy. B: Overlap of chronostratigraphic zones reflected in the lithostratigraphy.

previously been laid down, the limestone formation would overlap the shale formation in its turn. This can simply be regarded as a result of the sedimentary facies migrating at a different rate to the landward advance of the sea.

Effect of Topographic Relief.

Such features are typically associated with erosion surfaces which were horizontal, or nearly so. However, these features become more marked wherever overlap occurs against an unconformity which represents an erosion surface with considerable amount of topographic relief, as shown in Figure 7.8. Such an unconformity might be developed by a buried landscape, covered by continental deposits. These rocks are likely to be

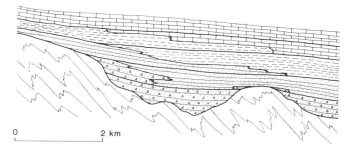

FIG. 7.8. Stratigraphic cross-section showing overlap associated with a buried erosion-surface of considerable relief.

deposited so that they are banked against the erosion surface at an appreciable angle. Thus, although these sediments are generally deposited with an initial dip, which might be quite high if they are scree deposits, it is always less than the slope of the erosion surface. This means that individual beds would wedge-out against the unconformity over relatively short distances, so that the overlap shown by the chronostratigraphy would be very marked. Furthermore, it is clear that the nature of the depositional environment is likely to change as the pre-existing topography becomes buried. Accordingly, there are likely to be lateral and vertical changes in sedimentary facies associated with such an unconformity, which would result in the marked overlap of stratigraphic formations against the unconformity. This means that overlap in such a case is defined by the lithostratigraphy as well as the chronostratigraphy.

Recognition of Overlap.

Further difficulties are introduced into recognising an unconformity from the outcrop pattern shown by a geological map wherever overlap affects the overlying rocks. It has already been described how overlap may be expressed by the lithostratigraphy in such a way that each formation forms a feather edge as it comes into contact with the underlying rocks along the unconformity. This is shown on a geological map wherever the boundaries between these formations intercept the boundary representing the unconformity at an angle, as it traced along the strike (see Fig. 7.9). This arises from the very nature of overlap, which requires that the bedding of the overlying rocks is not strictly parallel to the dip and strike of the unconformity at the base of the stratigraphic sequence.

To what extent this would obscure the nature of the outcrop pattern depends on the amount of topographic relief associated with the development of the unconformity as an erosion surface. If this is very slight, the overlap of individual formations against the unconformity would be very gradual, so that the outcrop of each formation would extend for a long distance along the unconformity before it disappeared. However, if the relief associated with an unconformity is greater, the overlap of individual formations against the unconformity is more marked, so that these formations wedge-out against the unconformity over a relatively short distance. This rather modifies the outcrop pattern because the boundaries between these formations would no longer be parallel, even approximtely, to the unconformity at the base of the stratigraphic sequence.

In effect, this means that the outcrop pattern associated with pronounced overlap is similar to that developed as a result of overstep. However, there is an important difference. It is the underlying formations which are truncated by the unconformity in the case of overstep, whereas it is the overlying formations which wedge-out against the unconformity as the result of overlap.

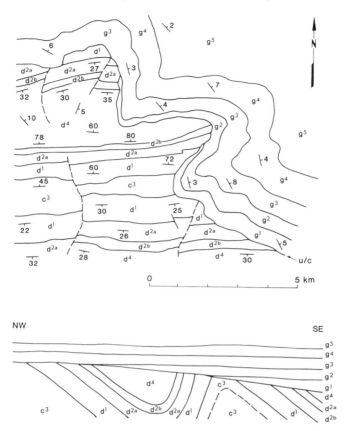

FIG. 7.9. Geological sketch-map and vertical cross-section showing the outcrop pattern developed as a result of overlap above an angular unconformity. Compare with Figure 7.4.

This means that a distinction can easily be made between overstep and overlap once the stratigraphic order of the two sequences has been determined. Thus, overstep occurs at the stratigraphic top of a sequence lying below an unconformity, whereas overlap would only be developed at the stratigraphic base of the overlying sequence, which becomes younger as it is traced away from the unconformity.

Unconformities and the Stratigraphic Column

Once the distinctive features of an unconformity have been recognised as a particular type of geological boundary, a stratigraphic column can be erected by studying the outcrop pattern shown by a geological map. This is formed by a number of stratigraphic sequences, each separated from one another by unconformities. Each sequence consists of a number of sedimentary and volcanic formations, placed in stratigraphic order according to the Principle of Superposition. The unconformities separating these sequences of sedimentary and volcanic formations from one another can be recognised by means of the overstep which occurs wherever the base of one stratigraphic sequence cuts across the structures developed in the underlying rocks. This allows these sequences to be distinguished from one another, while confirming that the stratigraphic order of each sequence has been correctly stated. It is then possible to place these sequences of sedimentary and volcanic formations in stratigraphic order, so drawing up a stratigraphic column. Once this has been done, a history of sedimentary deposition and volcanic activity, interrupted by intervals of uplift and erosion, can be inferred from the nature of the geological record.

The depositional episodes in this history would be marked by the accumulation of sedimentary formations, which were deposited chronologically in stratigraphic order, to form the conformable sequences of sedimentary rocks that constitute the stratigraphic column. A record of volcanic activity is preserved wherever the sedimentary formations within these sequences are replaced by volcanic rocks. It should be realised that all these formations are lithostratigraphic units, so that their boundaries might be diachronous. This introduces a difficulty since a diachronous formation would not have been deposited at the same time throughout the area where it now occurs. For example, the lowermost beds of such a formation could have been deposited at one place, while the uppermost beds of the underlying formation were still being deposited elsewhere.

This means in effect that a depositional history erected on the basis of stratigraphic order can only be applied strictly to a single locality within the area under consideration. However, any evidence for lateral changes in sedimentary facies should obviously be incorporated into such a depositional history. This is best done by describing the stratigraphic relationships which result from such changes, as defined by the wedging-out of individual formations within the stratigraphic sequence or the overlap of these formations at its stratigraphic base.

The intervals of uplift and erosion are marked by the unconformities which separate these sequences of sedimentary and volcanic formations from one another. This means that the stratigraphic record of sedimentary deposition and volcanic activity is interrupted after the deposition of each stratigraphic sequence by an interval of uplift and erosion. This represents a gap in the stratigraphic record. It corresponds to an interval of geological time which lacks any evidence for the deposition of sedimentary rocks.

However, it should be realised that sedimentary rocks might have been deposited during this interval of geological time, only to be removed from the stratigraphic record as a result of the erosion which accompanied the subsequent development of the unconformity at the base of the overlying sequence. This implies that the preservation of sedimentary rocks in the form of a stratigraphic record is incomplete. In particular, sedimentary rocks tend to be lost by erosion from the upper parts of stratigraphic sequences.

Contemporaneous and Posthumous Uplifts

The thickness of a stratigraphic sequence decreases to zero as it is traced to the margins of the sedimentary basin in which it was deposited. The area of non-deposition lying beyond the confines of such a sedimentary basin is known as a positive area, which did not subside during the development of the sedimentary basin. Indeed, if it had acted as the source for the clastic sediments deposited within the basin, there would be evidence that it had undergone uplift and erosion during this time. Such uplifts can be compared with the negative areas which develop into sedimentary basins as the result of long-continued subsidence.

It is commonly found that uplifts of various kinds separate sedimentary basins from one another. They are represented in the geological record by the angular unconformities which separate sedimentary sequences from one another. However, sedimentary sequences are commonly found to be affected by the earth movements after their deposition in such a way that they are preserved in structural basins. The limits of such a structural basin can be defined by the angular unconformity at the base of the sedimentary sequence now preserved within the basin.

Two interpretations can therefore be placed on such an unconformity where it forms the limits of a stratigraphic sequence against an area of older rocks, as shown in Figure 7.10. Firstly, it may have developed in response to uplift and erosion which occurred at the same time as deposition was proceeding elsewhere within the confines of a sedimentary basin. It then represents what is known as a contemporaneous uplift, flanking the sedimentary basins. This means that the sedimentary rocks underlying the unconformity are preserved as a result of differential subsidence, so that they never extended very far beyond the present limits of their outcrop. Secondly, such an unconformity may represent a posthumous uplift which affected the rocks deposited within a sedimentary basin after deposition came to an end. The sedimentary rocks preserved below such an unconformity would have originally extended far beyond their present outcrop. They are now found within the confines of a structural basin, formed as a result of later movements. Even although these rocks must have been deposited within a sedimentary basin of some sort, its form cannot be determined from their present mode of preservation. This can only be done if outliers of the same stratigraphic sequence are preserved from erosion over a much wider area.

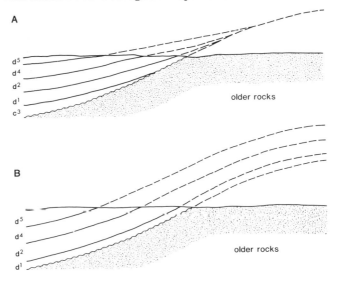

FIG. 7.10. Two possible interpretations of an unconformity as representing (A) a contemporaneous uplift and (B) a posthumous uplift. Note lack of overlapping relationships in Case B.

Use of Isopachyte and Lithofacies Maps.

The contemporaneous nature of an uplift can be established as the result of stratigraphic and sedimentological studies, which would allow the palaeo-geographic features of a particular region to be identified. Isopachyte maps show how the thickness of a particular stratigraphic unit varies throughout the area where it is now found. Such maps are contoured to show isopachytes as lines of equal stratigraphic thickness. Stratigraphic studies would be directed towards erecting a detailed chronostratigraphy for the sedimentary rocks that form such a unit. Any differences between this chronostratigraphy and the lithostratigraphy that has been established as the result of geological mapping could then be interpreted in terms of lateral changes in sedimentary facies. While isopachyte maps are used to show the varying thickness of each chronostratigraphic unit, which has been deposited during a particular interval of geological map, lithofacies maps can also be constructed to show how the lithological character of these units varies throughout the region under consideration. Such variations in thickness and lithology can obviously be used to determine the palaeogeography of the region, since they are directly related to this palaeogeography. This evidence is augmented by sedimentological studies establishing the source and provenance of the clastic sediments, the direction of sedimentary transport, and the varying nature of the depositional environments, which can again be related to the palaeogeography of the region.

Overlap and Marginal Unconformities.

It has already been emphasised that such evidence can rarely be abstracted from the study of a geological map. This means that the margins of a sedimentary basin are often difficult to recognise, unless the geological map shows that the stratigraphic formations thin to zero as they are traced towards an area of contemporaneous uplift. Such evidence might show the sedimentary basin to have a margin that was formed by a hinge which remained more-or-less stationary during the course of its development. However, it is commonly found that the sedimentary sequence deposited within the basins has a transgressive character, so that it encroaches on the area of uplift with the passing of geological time. This is commonly marked by the overlap of each stratigraphic formation in its turn, as it thins to zero against the margins of the sedimentary basin. Such evidence of overlap is extremely important in establishing the limits of a sedimentary basin. Since sedimentary basins often evolve by first expanding at the expense of the surrounding uplifts, and then by contracting as deposition comes to an end, it is often the case that their margins can only be located during the earlier stages in the deposition of a stratigraphic sequence. Even if there is no direct evidence of overlap, it can be inferred to take place wherever an uplift is covered transgressively by a sedimentary formation which elsewhere does not occur at the base of the stratigraphic sequence. Such a relationship must mean that the underlying formations are affected by overlap against the uplift.

Although stratigraphic formations may appear to be affected by overlap as they thin to zero against the expanding margins of a sedimentary basin, it must be admitted that similar features can be produced by the development of marginal unconformities, as shown in Figure 7.11. Such structures would be formed by oscillatory movements of transgression and regression, which occur in response to subsidence and uplift periodically affecting the margins of the sedimentary basin. The transgressive phase is marked by the deposition of sedimentary rocks over a wider area than where they are now preserved. The regressive phase is marked by the area of uplift expanding at the expense of the sedimentary basin in such a way that erosion takes the place of deposition. Renewed subsidence would then cause a local unconformity to be developed at the stratigraphic base of the overlying rocks.

This unconformity would cut across the underlying rocks so that they are overstepped towards the area of uplift, while it would disappear as it was traced in the opposite direction towards the sedimentary basin, where a more complete sequence of stratigraphic formations is likely to be preserved. However, the overstep associated with such an unconformity is difficult to distinguish from the overlap which would otherwise occur as the stratigraphic sequence thins to zero against the uplift. Such a distinction can often only be made if the unconformity cuts across more than one for-

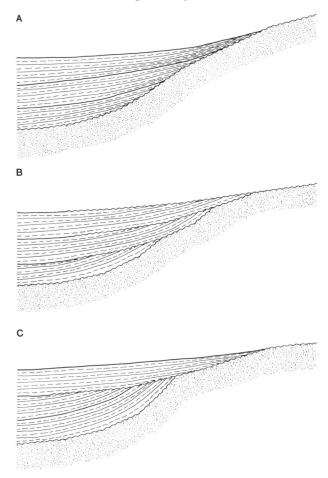

FIG. 7.11. Diagrams showing difficulty in distinguishing (A) overlap from (B) marginal unconformities unless (C) overstep is associated with a marginal unconformity.

mation within the underlying sequence. This can be contrasted with overlap, where the corresponding formation is only found in contact with one formation within the underlying sequence. Such a distinction cannot be made wherever a marginal unconformity cuts across only a single formation as it is traced towards the margin of the sedimentary basin. This means that the repeated overlap which is commonly shown by one formation after another at the margins of a sedimentary basin could be interpreted as a whole series of marginal unconformities, developed between these formations. It should be emphasised that all these relationships are developed between the individual formations within a stratigraphic sequence, which itself rests unconformably on top of older rocks forming the foundations of the sedimentary basin.

All this evidence of overlap, stratigraphic thinning and the development of marginal unconformities tends to be lost from the geological record as the result of uplift and erosion. This is likely to occur wherever a regression brings deposition to an end within the sedimentary basin. It is commonly found that the peripheral uplifts encroach on the sedimentary basin during such a regression, so that the sedimentary rocks deposited within the basin are only preserved in its centre. If the margins of the sedimentary basin cannot then be located from a study of the geological map, it can be assumed that the sedimentary rocks are preserved within a structural basin. The uplifts forming the present-day limits of such a basin would then be the result of posthumous movements, that occurred after the deposition of the sedimentary rocks now preserved within this basin. These rocks would be covered unconformably by a younger sequence of sedimentary rocks, overstepping the eroded margins of the sedimentary basin.

Use of Unconformities in Stratigraphic Dating

Although the development of an unconformity marks an important event in the geological history of a particular region, there are two distinct aspects to be considered in its interpretation. Thus, it marks a gap in the stratigraphic record, corresponding to an interval of geological time which lacks any evidence for the deposition of sedimentary rocks. It is quite likely that sedimentary rocks have been lost from the geological record as the result of uplift and erosion which occurred during this interval. However, the presence of an unconformity can also be used to date structural, igneous and metamorphic events as occurring during such a gap in the stratigraphic record. Although the dating of such events is based on the simplest of principles, they are fundamental to the development of geology as a historical science. These principles complement the Principle of Superposition since they allow a geological history to be inferred from the stratigraphic and structural features preserved by the rocks of the area under consideration. Such a history would embrace all the events which can be recognised from a study of the geological record.

Dating Folds and Faults.

The geological structures developed as a result of folding and faulting can first be considered. Folds can only be recognised from the nature of the outcrop pattern, as modified by the topography, taking any structural observations into account if they are plotted on the geological map to give the dip of bedding within these formations. Faults are usually shown in a distinctive way on a geological map, so that they cannot be mistaken for any other kind of geological boundary, unless a mistake has been made in the mapping.

Folds and faults can be dated stratigraphically wherever they are overstepped by sedimentary rocks lying above an unconformity, as shown, for example in Figure 7.9. They must be later than the sedimentary rocks which were deposited below the unconformity, since it is these rocks that are affected by the folding and faulting. However, they must be earlier than the sedimentary rocks which were deposited above the unconformity, since the folding and faulting does not affect these rocks. Such folds and faults are often known as pre-unconformity structures. They can be contrasted with the post-unconformity folds and faults which affect the sedimentary rocks deposited on top of the unconformity, as well as the underlying rocks. This serves to emphasise the way that an unconformity can be used as a stratigraphic marker to date structural features.

Although the dating of structural events in this way appears simple and straightforward, a difficulty is introduced wherever folding and faulting are renewed after the sedimentary rocks lying above an unconformity have been deposited, following an earlier episode of earth movements, as shown in Figure 7.12. This is generally the case wherever more than one unconformity is present, dividing up the stratigraphic column into a number of different sequences. The unconformities separating these sequences from one another represent intervals of geological time when folding and faulting have affected the pre-existing rocks. Accordingly, each sedimentary sequence is affected by all the episodes of folding and faulting which occurred after its deposition.

This means that each episode of folding and faulting in the geological history of an area has a cumulative effect, tending to accentuate the structures already present in the underlying rocks while causing new structures to be developed in the sedimentary rocks which have just been deposited above a particular unconformity. This would cause differences in structural style to be developed between the various sequences of sedimentary rocks, reflecting the nature of the folding and faulting that occurred during the various gaps in the stratigraphic record. As a result, it is possible to date folding and faulting as occurring after the deposition of one sequence, prior to the deposition of the next sequence in stratigraphic order, according to the differences in structural style which exist between the two sequences. These differences would be reflected in the outcrop pattern, which would show the underlying sequence to be more disturbed by the folding and faulting than the overlying rocks.

Such differences in structural style are also marked by the structural discordance which is developed between the two sequences of sedimentary rocks. This may be expressed by the overstep which is shown by the unconformity at the stratigraphic base of the overlying sequence, as it cuts across

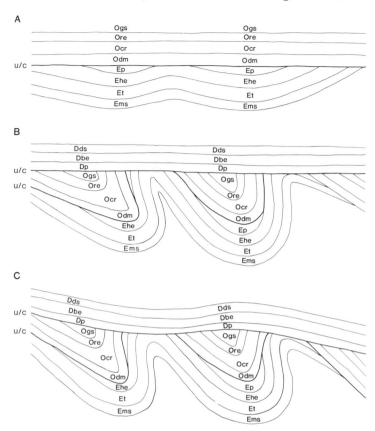

FIG. 7.12. Vertical cross-sections illustrating how the effects of earth movements on geological structures are cumulative in nature. A: Warping affects the Cambrian rocks prior to the deposition of the Ordovician sequence. B: Renewed earth-movements cause Cambrian and Ordovician rocks to be folded together, prior to the deposition of the Devonian sequence. C: Warping of the Devonian rocks affects the pre-existing structures developed in the underlying Cambrian and Ordovician rocks.

the pre-unconformity structures which are developed in the underlying rocks. Such overstep is only seen to advantage if the structural trends developed by the folding and faulting of these rocks are oblique to the outcrop of the unconformity between the two sequences. Under these circumstances, the nature of this overstep can be used to determine the original form of the pre-unconformity structures, which were developed prior to the deposition of the overlying rocks.

Dating Igneous Intrusions.

Geological maps commonly show the outcrops of igneous intrusions by means of distinctive colouring or ornament, while their petrological character is usually specified according to a particular scheme of lettering, as shown by the legend. This means that there should be no difficulty in recognising the outcrop of an igneous intrusion on a geological map, particularly if it is shown to cut across the structures developed in the surrounding country rocks.

Such an intrusion can only be dated accurately if it is overstepped, along with its aureole of thermal metamorphism in the country rocks, by sedimentary rocks lying above an unconformity. It must then have occurred after the deposition of the sedimentary rocks which now form its country rocks, since an igneous intrusion is always younger than the sedimentary rocks which it intrudes. However, the intrusion must also have taken place before the deposition of the sedimentary rocks which are now found above the unconformity, since this represents a surface of erosion cutting across the igneous body.

This is usually taken to imply that an igneous intrusion can generally be dated as occurring after the deposition of the sedimentary sequence which it intrudes, prior to the deposition of any sedimentary rocks which overlie this sequence unconformably. The intrusion would then have occurred during the gap in the stratigraphic record as represented by the unconformity between the two sequences. However, this may not be the case, since it is only possible to date an igneous intrusion as younger than the sedimentary formations which it is seen to intrude at the present level of erosion. There is no reason to suppose that it must be younger than the uppermost beds of the stratigraphic sequence which it intrudes, although this is generally assumed unless there is any evidence to the contrary.

In general, an igneous intrusion may only be dated as occurring during the deposition of a sedimentary sequence if it is associated with the eruption of volcanic rocks at the earth's surface, which then become incorporated into this sequence. There is rarely any direct evidence to date such an intrusion, unless it forms an obvious part of a volcanic edifice, since it is hardly ever possible to trace an igneous intrusion upwards into an extrusive lava-flow or pyroclastic deposit. However, circumstantial evidence may point to such an association. For example, a series of igneous intrusions might be formed by rocks which are similar in composition to the products of volcanic activity, bearing in mind the contrast in their mode of occurrence. If these intrusions were found to cut only the sedimentary rocks lying stratigraphically below the corresponding horizon of volcanic rocks, it would be reasonable to conclude that these groups of intrusive and extrusive rocks were contemporaneous with one another. Otherwise, it is usually assumed that such intrusions were formed after the whole of the sedimentary sequence had been deposited.

There is another problem in the stratigraphic dating of igneous intrusions, which needs to be mentioned. Thus, it is only possible to show that these intrusions are older than a particular sequence of sedimentary rocks if they are seen to be overstepped. However, it is commonly the case that such an intrusion is nowhere found in contact with the sedimentary rocks of the overlying sequence. This means that there is a distinct possibility that it could have been formed after the deposition of these rocks. Again, it is only circumstantial evidence which can be used to decide against such an interpretation. For example, a particular intrusion may be associated with a number of other intrusions, none of which are seen to intrude the sedimentary rocks of the overlying sequence. This would suggest that all these intrusions were formed prior to the deposition of these rocks. This interpretation would be strengthened if it could be shown that some of these intrusions at least were overstepped by the unconformity at the stratigraphic base of this sequence. Indeed, a particular intrusion could be proved to be older than these rocks if it was cut by another intrusion, which was itself overstepped by the unconformity at the stratigraphic base of the overlying sequence.

It should be realised that other evidence may be used to date an igneous intrusion apart from its structural and stratigraphic relationships. For example, if it is composed of a sufficiently distinctive rock type, it may be possible to identify any detritus formed by its weathering that has been incorporated as fragmental material within the overlying sequence. Since this detritus could only be formed through the unroofing of the igneous intrusion as the result of uplift and erosion, its presence is clear evidence that the igneous intrusion occurred prior to the deposition of the sedimentary rocks which now contain its detritus. Such evidence would not be shown on a geological map, even although it might well be mentioned in the legend.

Dating Regional Metamorphism.

Unconformities can also be used to date episodes of regional metamorphism, which affect sedimentary rocks after their deposition. This can be done wherever sedimentary rocks, altered by regional metamorphism, are overstepped by a younger sequence of sedimentary rocks lying above an unconformity, which are not so affected. Such a relationship implies that the deposition of the sedimentary rocks which are now found below the unconformity was followed by an episode of regional metamorphism, after which the sedimentary rocks lying above the unconformity were laid down. Since regional metamorphism only occurs at considerable depths under a particular geothermal gradient, where temperatures are sufficiently high for metamorphic reactions to take place, the accumulation of sedimentary rocks must have continued until enough overburden was present for regional metamorphism to affect the rocks now lying immediately below the unconformity. This means that a considerable amount of uplift and erosion must have occurred after the regional metamorphism came to an end, in order to remove this thickness of overburden. Such an unconformity must therefore represent a relatively large break in the stratigraphic record corresponding to an appreciable thickness of sedimentary rocks, once deposited as part of the underlying sequence and now lost by erosion.

Although it is possible that regional metamorphism could have started at depth while the deposition of sedimentary rocks continued at the earth's surface, there is rarely any direct evidence to support such an interpretation. Indeed, the mineral assemblages preserved in metamorphic rocks usually reflect nearly the highest temperatures reached in their evolution from sedimentary rocks. Even if the geothermal gradient varies during the course of regional metamorphism, such temperatures are mostly likely to be reached once the overburden has attained its maximum thickness after the end of deposition. This conclusion is also supported by the fact that the climax of regional metamorphism commonly occurs after a considerable amount of folding and deformation has affected the rocks, which is unlikely to occur while deposition is still proceeding at the surface. Accordingly, it is usually assumed that sedimentary rocks undergo regional metamorphism after deposition has come to an end so that it may be regarded without any qualification as a later event.

Unconformities as Gaps in the Stratigraphic Record.

The dating of structural, igneous and metamorphic events in this way implies that they occurred during a particular gap in the stratigraphic record, as represented by an unconformity. As it stands, this gap corresponds to the difference in stratigraphic age between the rocks separated by the unconformity. Thus, the stratigraphic record appears to end with the deposition of the youngest beds which are now preserved from erosion below the unconformity, while it definitely resumes at a later date with the deposition of the oldest beds which are now found above the unconformity. However, this gap in the stratigraphic record is marked by uplift and erosion. It is almost inevitable that part of the underlying sequence will be lost by erosion. This implies that even younger rocks were deposited conformably as part of the stratigraphic sequence which is now found below the unconformity, only to be lost as a result of the erosion accompanying its formation.

Accordingly, the present gap in the stratigraphic record would represent a longer interval of geological time than that which originally elapsed between the cessation of sedimentation, prior to the earth movements, and its resumption after uplift and erosion had taken place. In fact, it is not easy to determine when sedimentation ceased before the development of an unconformity, unless the interval of geological time represented by the unconformity is very short. It could have ceased at any time after the deposition of the youngest rocks which are now found below the unconformity, prior to the deposition of the oldest rocks lying above the unconformity. This obviously introduces a further element of uncertainty into the stratigraphic dating of structural, igneous and metamorphic events. In contrast, the resumption of sedimentation after uplift and erosion have taken place can be accurately dated if there is sufficient evidence to determine the stratigraphic age of the sedimentary rocks which occur immediately above the unconformity.

Rejuvenation of Folds and Faults

A problem in the stratigraphic dating of folds and faults may be encountered wherever folding and faulting are renewed after the sedimentary rocks lying above an unconformity have been deposited, following an earlier episode of earth movements. This would arise if the renewed folding and faulting modified the structures already present in the sedimentary rocks below the unconformity, rather than causing any new folds or faults to be developed in these rocks. Such structures are said to undergo rejuvenation, whereby their original character becomes altered by the renewed folding and faulting. It is obvious that such structures cannot be assigned to a single episode in the geological history of the area. However, the nature of the renewed folding and faulting can be judged from their effect on the sedimentary rocks which are now preserved above the unconformity. It might then be possible to determine how the structures in the underlying rocks developed in response to the two episodes of folding and faulting which can be recognised from the geological record. This would allow the different stages involved in the development of these structures to be dated stratigraphically. This principle can obviously be extended, assuming that several episodes of folding or faulting were responsible for the development of a particular structure.

Evidence for Rejuvenation of Folds.

It must be admitted that folds rarely show any clear evidence of rejuvenation. However, the presence of folded rocks below an unconformity must have some effect on any subseuqent folding that tends to occur. Much depends on the style and relative orientation of the structures affecting the underlying rocks. If they were relatively open folds trending approximately at right angles to the later compression, it is likely that these folds will be tightened by the later folding. This would cause the overlying rocks to be folded along the same lines as the structures already developed in the rocks lying below the unconformity in such a way that the two sets of fold structures would correspond very closely to one another, as shown in Figure 7.13. The unconformity would then overstep a series of progressively older rocks as it is traced away from the axial traces of any synclines that are

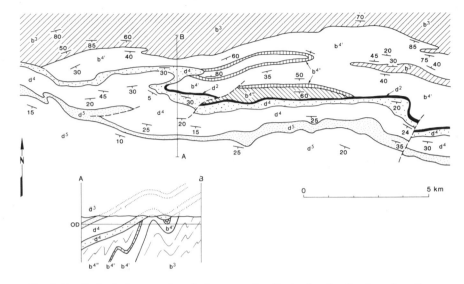

Fig. 7.13. Geological map and vertical cross-section illustrating the situation where the structures developed above and below an unconformity have the same trend. (Redrawn with modifications from Sheet 228 of the Geological Survey of England and Wales.)

developed in the overlying rocks, towards the intervening anticlines. If this is found to be the case, the folds already present in the sedimentary rocks below the unconformity must have been accentuated by the folding that occurred after the overlying rocks were deposited, so that the pre-unconformity structures were tightened as a result of the post-unconformity folding.

The effects of the post-unconformity folding can be removed from the geological record by restoring the bedding of the overlying rocks to a horizontal position, assuming that these rocks were originally deposited with little or no dip, as shown in Figure 7.12. How this should be done depends on the nature of the post-unconformity folding. Even if this can be determined, which may not be easy to do, it is often difficult to decide how these movements modified the pre-unconformity structure of the underlying rocks.

Under these circumstances, it is usually assumed that these rocks were affected by simple tilting without any internal deformation as a result of the post-unconformity folding. This means that the angular discordance between these rocks and the overlying sequence remains the same despite the later movements. This assumption is usually followed in any attempt to reconstruct the structural features developed in the underlying rocks, prior to the deposition of the overlying sequence of sedimentary rocks. This is termed a palinspastic reconstruction of the geological structure, as it existed prior to the formation of the unconformity. It should be realised that such a palinspastic reconstruction can only be approximately correct if it is constructed on the basis of this assumption.

Complications are likely to be introduced wherever the later compression occurs at an oblique angle to the earlier folding. Even although the structures developed in the underlying rocks may be so orientated that they can tighten in response to the later compression, this may need to be accompanied by the development of complex fault-patterns in order to acommodate the overall deformation. The overlying rocks would then be affected by equally complex patterns of folding and faulting, even although the folds would be more open structures than those developed below the unconformity. Such patterns would be controlled by the nature of the geological structures originally developed in the rocks underlying the unconformity.

Development of Interference Patterns.

If the folds occurring below an unconformity do not have an orientation which would allow them to respond to a later compression by tightening, an entirely new set of fold structures may be produced. Since the two sets of fold structures would differ in trend, their intersection will produce interference patterns in the rocks underlying the unconformity. Typically, culminations and depressions are formed along the hinge lines of the earlier folds, so converting these folds into domes and basins. Domes are formed wherever two anticlines cross one another, while basins are produced where two synclines cross one another. The saddles between these structures mark the points where an early anticline is crossed by a late syncline, or where an early syncline is crossed by a later anticline. The stresses produced by the renewed folding of already-folded rocks may cause a complex pattern of faults to be developed in these rocks. It should be emphasised that domes and basins can also be formed in response to a single episode of folding. It can often only be proved that such structures were produced by the interference of two sets of folds, if one set is seen to affect an overlying sequence of sedimentary rocks while the other set is overstepped by the unconformity at the base of this stratigraphic sequence. Such relationships are not very common in the geological record.

Differential Folding and Faulting.

The rejuvenation of pre-existing folds, or the development of interference patterns, requires that the rocks underlying an unconformity have a structure which is able to register the effects of any subsequent folding. This means in effect that the folds originally present in these rocks must have been relatively open structures, which can be affected by the renewed folding. Otherwise, this episode may have little or no apparent effect on the structure of the rocks which have already been folded below the unconformity, even although the overlying rocks are folded to a considerable extent. For example, it is commonly found that the effects of subsequent folding are apparently lost on the rocks lying below an unconformity where they have previously been tightly folded on upright axial planes. However, if the later deformation is sufficiently intense, the underlying rocks may well deform by fracture and shearing to produce a complex array of closely spaced fault planes. The movements on these planes could then be accommodated upwards by the folding of the rocks above the unconformity, as shown in Figure 7.14. Again, it is likely that the renewed deformation will take advantage of pre-existing planes of weakness in the underlying rocks, so controlling the nature of the structures developed above the unconformity, as shown in Figure 7.14.

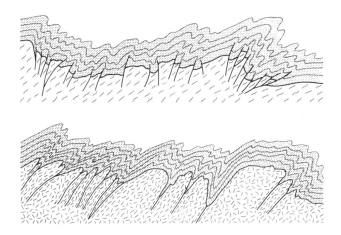

Fɪɢ. 7.14. Vertical cross-sections illustrating differences in structural style across an unconformity.

Evidence for Rejuvenation of Faults.

By way of contrast, faults are more likely to undergo rejuvenation since they form pre-existing planes of weakness along which renewed movements may well take place. However, it is very difficult to determine the nature of these movements wihtout making any special assumptions. This problem has been rather neglected in the literature so that it is worth emphasising its importance. It is obvious that any pre-unconformity faulting would produce a fault which did not extend upwards to affect the sedimentary rocks above the unconformity. However, if post-unconformity faulting occurred so that this fault was rejuvenated, it would extend upwards to affect the overlying rocks. It therefore needs to be considered whether or not such a structure differs in any way from a fault developed entirely in response to post-unconformity movements, which would also affect the underlying rocks.

Consider first the traces made on the opposite walls of the fault plane by the unconformity (U/C) and an underlying horizon P, assuming that neither the unconformity nor the underlying horizon have been folded, as shown in Figure 7.15. Each structure maintains the same dip and strike on either side of the fault plane. This means that the traces corresponding to one another on the opposite walls of the fault plane would be parallel. Together, these traces define a distorted T-shape, which is displaced across the fault plane.

This displacement could have been accomplished in a single step, as shown by the line *AD* in Figure 7.15. The faulting would then have occurred after the sedimentary rocks had been deposited above the unconformity, so that it would be entirely post-unconformity in age. If so, *A* and *D* mark the two points on the underlying horizon *P* which were originally coincident with one another. However, it might be assumed that this displacement had taken place in two steps. For example, the first step could have involved a pre-unconformity displacement along the line *AB*, which affects only the underlying horizon. Uplift and erosion then produced the unconformity which acted as the stratigraphic base for the subsequent deposition of sedimentary rocks, so that it cuts across the fault in the underlying rocks.

Renewed movements on this fault are then required in order to develop the final geometry as shown in Figure 7.15C. These movements involve a post-unconformity displacement along the line *BC*, affecting the underlying horizon. The corresponding displacement which affects the unconformity itself can be designated by the line *B'D*. This means that *A* and *C* now mark the two points on the underlying horizon on either side of the fault plane, which were originally coincident with one another. This disparity arises because the upthrown block undergoes more erosion than the down-faulted block, prior to the deposition of the overlying rocks. The intersection of the unconformity with an underlying horizon is not an unique point, which can be matched across the fault plane in order to determine the net displacement.

In fact, we made an arbitary assumption that the first displacement took place along a particular line *AB* when it could have occurred along any other line within the fault plane. The nature of the second displacement along the line *BC* varies according to the assumption made about the first displacement, since it must occur in such a way that the present geometry can be developed across the fault plane. In other words, the point *C* would lie anywhere along the trace made by the underlying horizon against the hanging wall of the fault plane, depending on the nature of the first displacement.

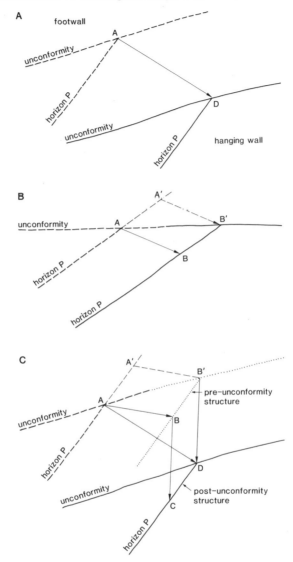

Fig. 7.15. A: Diagram illustrating the traces made by an unconformity and an underlying horizon P on the footwall (dashed lines) and hanging wall (solid lines) of a fault. Note that the net displacement AD could be accomplished so that the points A and B on either side of the fault plane are equivalent to one another. B: Pre-unconformity faulting results in a net displacement along the line AB. Note that the trace on the hanging wall reaches the surface of erosion now represented by the unconformity at the point B′. C: Post-unconformity faulting results in a net displacement along the line B′D. Note that the points originally coincident with A and A′ on the footwall trace out the paths ABC and A′B′D respectively, so that the points A and C now correspond to one another on either side of the fault plane.

It should be emphasised that vertical cross-sections only show the structure of faulted rocks in two dimensions, so that it tends to be assumed that
all the faulting has occurred as a result of dip-slip movements within this
plane. Consider, for example, the vertical cross-section shown in Figure
7.16, which appears to indicate that the fault has been rejuvenated. Thus, if

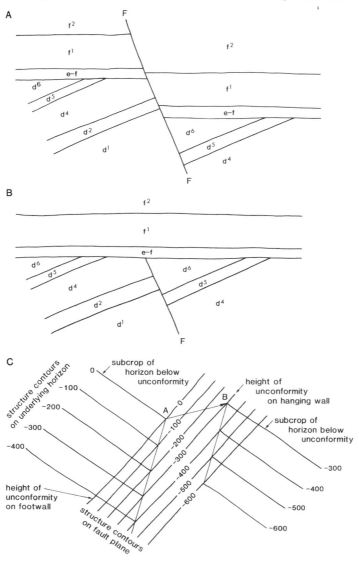

FIG. 7.16. A and B: Apparent evidence of pre-unconformity faulting as shown by a
vertical cross-section. C: Structural plan showing how the displacement as seen in
cross-section can be accomplished by a single episode of oblique-slip movements along
the line AB. The unconformity is assumed to be horizontal.

the unconformity is restored to such a position that it regains its original continuity across the fault plane, the undelying horizon still appears to be faulted. This might be taken as evidence for pre-unconformity movements. However, this conclusion is based on the assumption that the post-unconformity faulting occurred entirely as a result of dip-slip movements in the plane of the vertical cross-section. If strike-slip movements also took place during this episode, they could restore the continuity of the underlying horizon, provided that its strike was oblique to the trend of the fault plane. Only if the underlying horizon has the same strike as the fault plane, so that it would not be affected by any strike-slip movements, would it be necessary to introduce an episode of pre-unconformity faulting to account for the development of the faulted structure. Although it may well be the case in general that such an episode of pre-unconformity faulting did occur, this cannot be established by means of the evidence shown by the vertical cross-section.

Evidence for Pre-unconformity Faulting.

The problem as stated therefore provides no evidence to determine whether or not a fault cutting an unconformity has undergone any re-juvenation. However, it may be possible to determine the net displacement, as far as it affects the rocks lying either above or below the unconformity. The structural and stratigraphic features that can be used for this purpose have already been listed in Chapter 5. Two particular features will be chosen, simply as examples, to show what further information is provided by this evidence.

The net displacement affecting the rocks lying above the unconformity can be found wherever any overlap occurs at the base of this stratigraphic sequence, as shown in Figure 7.17. The feather edge of an overlapped formation can then be matched across the fault plane, so defining the line $B'D$. By reversing the post-unconformity displacement given by this line, it can definitely be established that the underlying rocks were affected by pre-unconformity faulting, wherever they still have some separation across the fault plane. The nature of the pre-unconformity faulting cannot be determined from the evidence shown by Figure 7.18. Strictly speaking, the opposite conclusion cannot be reached, even if the underlying rocks do not show any separation across the fault plane. Under these circumstances, unconformity faulting might well have taken place if the net displacement occurred parallel to the traces made by an underlying horizon against the fault plane. Since the underlying rocks would then show no separation across the fault plane, the absence of such separation cannot be taken as evidence that pre-unconformity faulting did not occur.

Likewise, the net displacement affecting the rocks lying below the unconformity can be determined wherever it is possible to match a fold hinge

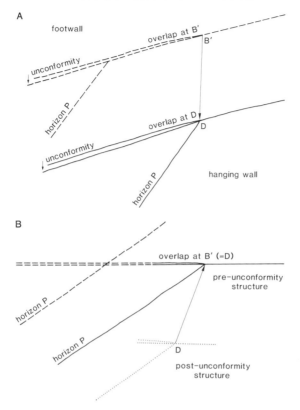

Fig. 7.17. Evidence for pre-unconformity faulting. A: Traces made by an unconformity showing overlap at its base, and an underlying horizon P, against the walls of a fault plane. B; Removing the effects of post-unconformity faulting by matching up the feather edges resulting from overlap at the points B′ and D shows that pre-unconformity faulting did occur although its nature cannot be determined.

across the fault plane. This then defines the line *AF*, which gives the total displacement as far as the underlying rocks are concerned, as shown in Figure 7.18. By reversing this displacement, it is possible to determine if the rocks lying above the unconformity can be restored to their original continuity. If so, it may be assumed that no pre-unconformity faulting has taken place, unless post-unconformity faulting has occrred with a net displacement parallel to the traces made by the unconformity against the fault plane. If the overlying rocks cannot be restored to their original continuity, the conclusion can definitely be reached that pre-unconformity faulting did occur. However, the exact nature of either set of fault movements cannot be determined from the evidence shown in Figure 7.18, where the line *AE* is purely arbitrary.

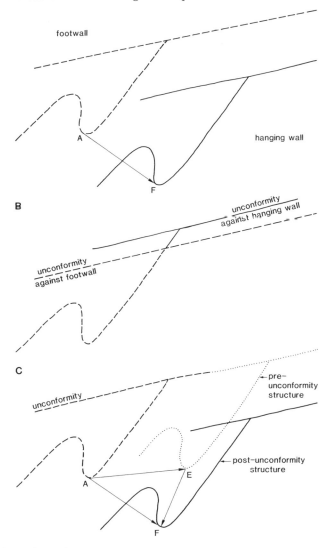

Fig. 7.18. Evidence for post-unconformity faulting. A: Traces made by an unconformity above a folded horizon against the walls of a fault plane. B: Removing the overall effects of the faulting on the underlying rocks by matching up the fold hinges at A and F shows that post-unconformity faulting did occur. C: Diagram showing how present structure might have developed by pre-unconformity and post-unconformity displacements along the lines AE and EF, respectively. Note that E is an arbitary point, so that the nature of these episodes of faulting cannot be determined.

Thus, although it is possible to establish that pre-unconformably faulting did occur according to this evidence in a particular case, any conclusion to the contrary cannot be proved unless it is possible to determine the net

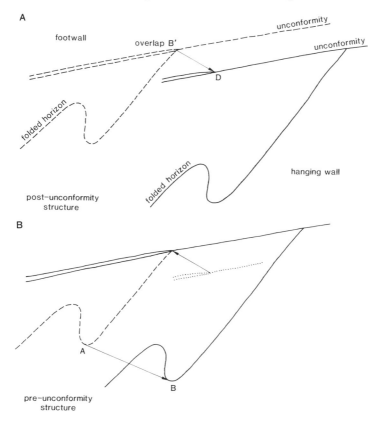

FIG. 7.19. Evidence allowing both pre-unconformity and post-unconformity displacements to be determined for a fault plane. A: Present structure. B: Pre-unconformity structure found by removing the effect of post-unconformity displacement along the line B′D.

displacement for both the rocks lying above, and lying below, the unconformity. This allows the lines *AB* and *B′D* to be determined in Figure 7.15.

The sort of evidence that can be used is shown in Figure 7.19 where overlap associated with the unconformity and a fold hinge in the underlying rocks allows both pre-unconformity and post-unconformity displacements to be determined. However, as emphasised in Chapter 5, it is often only circumstantial evidence which is available to establish the nature of the faulting. In particular, the faults affecting the rocks above an unconformity may have a character which would allow them to be recognised as dip-slip rather than strike-slip structures. It would then be possible to determine whether or not any pre-unconformity faulting had taken place. The absence of such evidence means that it is often difficult if not impossible to date stratigraphically the various stages involved in the development of a fault as a rejuvenated structure.

Palaeogeological and Subcrop Maps

The geometrical features of an angular unconformity can be studied through the use of structure contours. For convenience, assume that these contours are drawn on a planar surface, dipping uniformly in a particular direction. This surface can then be contoured to form an equally spaced set of straight and parallel lines. To represent an unconformity, one set would be drawn on the unconformity itself, while another set could be drawn of any suitable horizon within the underlying formations. The two sets of contour lines are shown in Figure 7.20.

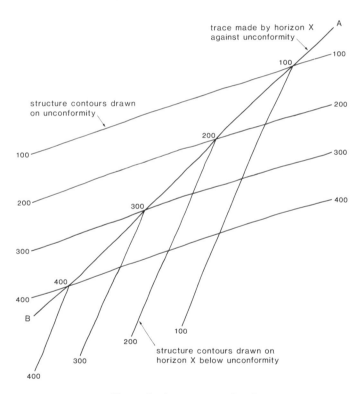

FIG. 7.20. Structure contours illustrating how an unconformity cuts across an underlying horizon (X) along a line AB.

To find how an underlying horizon meets the unconformity from underneath it is only necessary to consider the structure contours drawn at equivalent heights on the two surfaces. Taken in pairs, these contour lines will intersect one another in a series of points which lie along a straight line *AB*. This line corresponds to the intercept made by the underlying horizon on the surface of the unconformity, so that it represents the trace made by this horizon against the unconformity. Note that the contour lines drawn on this horizon have not been continued beyond this line, where it ends against the unconformity.

Subcrop Maps.

The trace made by such a horizon against an unconformity represents the line across which the unconformity oversteps from one formation to another within the underlying sequence. In fact, this line marks a geological boundary between two underlying formations, occurring on either side of this horizon, as it would be mapped beneath the unconformity at the base of the overlying sequence. Obviously, all the other horizons separating the various formations from one another within the underlying sequence could be mapped in the same way, by considering how each horizon in turn intercepts the unconformity. The traces made by all these horizons would then define the "outcrop" of the intervening formations in contact with the unconformity, as shown in Figure 7.21.

This would be a type of geological map. It is termed a subcrop map because it is produced by the "outcrop" of particular formations below a surface, as represented in this case by an unconformity. The term "subcrop" is introduced as a shortened form of the phrase "subsurface outcrop", meaning the outcrop which occurs below the earth's surface against a buried unconformity, for example. Such a subsurface outcrop would be seen if the unconformity was stripped of its overlying rocks. Subcrop maps are important in the present context because they show how the rocks lying above an unconformity overstep the underlying formations throughout a particular area, thus allowing the overstep associated with the unconformity to be mapped.

Palaeogeological Maps.

Unless the bedding of the rocks lying above the unconformity is still horizontal, it can be assumed that these rocks have been faulted and folded, after they were laid down. If the effects of these movements can be removed in some way, the unconformity would regain its original flat-lying position. This is known as a palinspastic reconstruction of the unconformity. Figure 7.12 shows a palinspastic reconstruction of a vertical cross-section, in which the effects of faulting and folding on the rocks lying above an unconformity have been removed. The subcrop map for such an unconformity can be modified in the same way, to show how this unconformity overstepped the underlying rocks when the overlying rocks were horizontal.

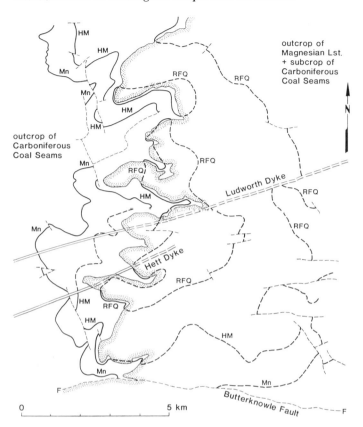

FIG. 7.21. Geological map showing the subcrop traces made by Upper Carboniferous coal seams below the Permian Magnesian Limestone of Country Durham, NE England. Coal seams are Main (Mn), High Main (HM) and Ryhope Five-Quarter (RFQ), given in ascending order. (Map redrawn with modifications from Sheet 27 of the Geological Survey of England and Wales.)

Since an unconformity is formed by an erosion surface, it must represent the earth's surface as developed at some time in the geological past, immediately before the rocks lying above the unconformity were deposited. The subsurface outcrop of the underlying formations against this unconformity then defines a geological map as it would have existed at that time. Accordingly, such a map is known as a palaeogeological map, corresponding to a time just earlier than the stratigraphic age of the rocks forming the base of the overlying sequence. This means that a palaeogeological map is a palinspastic reconstruction of a subcrop map. However, if the rocks lying above the unconformity are not greatly affected by folding and faulting, there would be little difference between the two types of geological map. Accordingly, subcrop maps are sometimes known as palaeogeological maps, even although a strict distinction really ought to be drawn.

Palaeogeological maps can be interpreted in the same way as geological maps. However, unless they are constructed for an unconformity which represents a buried land-surface with considerable amount of topographic relief, they are much easier to understand. Under these circumstances, a palaeogeological map can be interpreted directly in terms of geological structure. Any folding or faulting which affected the rocks lying below an unconformity before the deposition of the overlying rocks can be recognised from the outcrop pattern, without the necessity of taking the effects of topographic relief into account. This gives the pre-unconformity structure of the area covered by the palaeogeological map, since it is defined by the folds and faults which are truncated by the unconformity as a horizontal surface of erosion.

Accordingly, the pre-unconformity strike of the underlying formations is given directly by the trend of the formation boundaries across the palaeogeological map. Likewise, the pre-unconformity dip of these formations can be found directly from the width of their outcrops, assuming that the stratigraphic thickness of each formation is known. It would generally be assumed that these formations dipped towards the younger rocks, unless any evidence for overturning was available. Anticlinal areas would be recognised by the outcrop of older rocks, while synclinal areas would be marked by the outcrop of younger rocks. Faults could be recognised wherever a structural discontinuity was developed in the outcrop pattern. Further complications might be developed if the rocks lying below the unconformity formed more than one stratigraphic sequence, so that the younger formations overstepped a set of older formations along a geological boundary which could then be recognised as an unconformity. Finally, palaeogelogical maps would show igneous intrusions wherever they have been unroofed by erosion, prior to the deposition of the overlying rocks.

Mapping Subsurface Outcrops.

Subcrop and palaeogeological maps can rarely be constructed directly from the information provided by a geological map. Instead, they are usually prepared from subsurface data, which has been obtained by drilling bore holes in areas of economic interest. The only exception to this statement concerns those areas where the rocks lying above an unconformity are highly irregular in their outcrop, so that they form a series of outliers in advance of the main outcrop, which might also be interspersed with inliers of the underlying rocks. A subcrop map showing the traces made by the underlying formations against the unconformity could then be drawn for those areas where the younger rocks are preserved. This can be compared with a geological map which would show the boundaries between these formations wherever they outcrop at the earth's surface, away from the areas underlain by younger rocks.

It is obvious that these two maps must correspond to one another where they meet along the geological boundary representing the unconformity between the two sequences, as shown, for example, by Figure 7.21. Accordingly, any formation boundary shown on the subcrop map must end against the unconformity at those points where the unconformity cuts across the corresponding horizon on the geological map. This allows a subcrop map to be constructed from a geological map by drawing lines through all those points where the unconformity cuts across particular horizons within the underlying rocks. These should be drawn as solid lines where they cross the outcrop of the youger rocks, since this is the area where the subsurface outcrop of the underlying formations can be mapped against the unconformity itself. However, they can be continued as dashed lines across the intervening areas, where these formations are exposed at the earth's surface. This can be taken to indicate that the formation boundaries shown on the subcrop map are mere projections above the earth's surface. Thus, an unconformity can be projected beyond its outcrop, above the present level of erosion, to show how it would intersect the various formations within the underlying sequence, which have now been removed as a result of this erosion.

This kind of extrapolation allows a subcrop map to be extended over a whole region. However, it is important to realise that the formation boundaries so projected on a subcrop map and the geological boundaries between these formations at the present level of erosion, as shown on a geological map, do not correspond to one another under normal circumstances. They would only do so if the bedding of the underlying rocks was vertical or if the unconformity at the base of the overlying rocks was very close to the present level of erosion. Otherwise, a correction needs to be made for the effect of the topography on the outcrop of the underlying formations, below the level of the unconformity.

Structural Cross-sections through Unconformities

Strictly speaking, the cross-section of an angular unconformity should be drawn in an inclined plane lying at right angles to the direction in which the unconformity intersects the bedding of the underlying rocks, since the angle between any two planes is always measured in a plane at right angles to their intersection. Such a cross-section through an unconformity would then show the true discordance between the two sequences, and their correct stratigraphic thickness. Such an inclined cross-section cuts obliquely across the strike of both sequences, except where the subsurface outrops of the underlying sequence are parallel to either the dip or the strike of the overlying rocks. Even if it were drawn in a vertical plane, apparent dips would have to be used in its construction. As it is drawn in an inclined plane, a further correction needs to be introduced. Although this can be done using structure contours, the method is rather cumbersome. Such difficulties are avoided by using the two methods now to be described for the construction of a cross-section through an angular unconformity.

Cross-sections Along the Strike.

The first construction is based on the principle that, just as folds can be viewed down-plunge, unconformities can be viewed down-dip. However, it can only be applied if the line *AB* marking the intersection between the unconformity and an underlying horizon in Figure 7.20 is perpendicular to the strike of the overlying rocks, so that it is parallel to their dip. This only occurs if the underlying horizon has an apparent dip at right angles to the strike of the unconformity, equal to the true dip of the overlying rocks. The outcrop pattern then provides a distorted view of the unconformity, which can be corrected simply by viewing the geological map at an oblique angle, parallel to the true dip of the overlying rocks. Such a foreshortened view of an unconformity shows the overlying rocks to be horizontal, while providing a true cross-section of the rocks lying below the unconformity, as constructued in a plane at right angles to the intersection made between the unconformity and the bedding of these rocks (see Fig. 7.22).

This means that the same method as used to construct the profile of a plunging fold can be employed to construct the true cross-section of such an unconformity. It will be assumed that the topography has no effect on the outcrop pattern. A square grid of straight lines is then drawn across the geological map so that one set of grid lines is parallel to the strike of the rocks lying above the unconformity. Points are marked off where the various boundaries shown on the geological map across this set of grid lines. A rectangular grid of straight lines is then constructed on a separate diagram. One set of grid lines is drawn vertically so that they maintain the same spacing, while the other set of grid lines is drawn horizontally in such a way that their spacing is reduced by sin α, where the angle α gives the true dip of the rocks lying above the unconformity. The points marked on the geological map are then transferred to equivalent points on the horizontal grid lines. The geological boundaries are drawn through these points, using the grid lines as a framework so that a true cross-section of the rocks lying below the unconformity can be constructed. Finally, the cross-section is completed by drawing in the rocks lying above the unconformity as horizontal layers with a true thickness which is found by reducing the width of their outcrop by an amount equal to sin α.

This construction can only be used to give a true cross-section if the intersection made with an underlying horizon plunges down the dip of the unconformity. However, it may provide a distorted view of the unconformity,

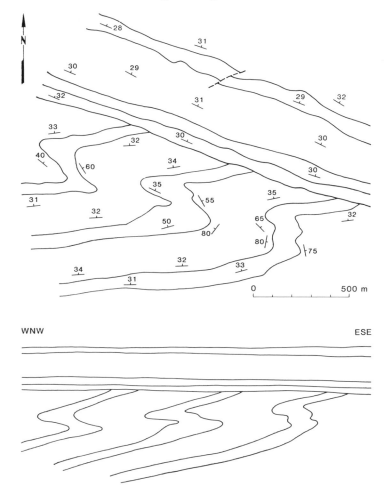

Fig. 7.22. Geological map and structural cross-section showing an unconformity corresponding to the situation discussed in the text where the subcrop of the underlying beds against the unconformity trends down the dip of the overlying beds to the NNE. Note that the folds affecting the underlying beds plunge at 30° towards the NNE, so allowing the true structure to be shown in profile on the cross-section.

even if this condition is not met. Consider, for example, the general case where the underlying formations outcrop at an acute angle to the unconformity. A foreshortened view of the geological map, as observed down-dip, would show these formations to be inclined in a certain direction. This corresponds approximately to the dip of the bedding, as seen in cross-section, wherever these formations dip towards the unconformity. Younger formations would then appear in this direction, assuming that no overturning has taken place. Although such a down-dip view of an unconformity does not truly represent the angular relationships between the two sequences, it does display the salient features of the structures affecting the underlying sequence, as truncated by the unconformity itself. This means that a diagrammatic cross-section of an unconformity may be obtained simply by viewing the geological map at an oblique angle, parallel to the dip of the rocks lying above unconformity, wherever the underlying rocks dip towards the uncoformity. However, this method cannot be applied if these rocks dip away from the unconformity, since they would then appear to be inclined in the wrong direction.

Vertical Cross-sections.

The second construction is based on the methods which are normally applied to the drawing of structural cross-sections. Strictly speaking, such methods can only be applied if the line *AB* marking the intersection between the unconformity and an underlying horizon in Figure 7.20 is parallel to the strike of the overlying rocks. This can only occur if the rocks separated from one another by the unconformity have the same strike. This is the case, for example, if post-unconformity tilting had taken place about the same horizontal axis as pre-unconformity folding, so that the structural trends developed on either side of the unconformity are parallel to one another. A true cross-section can then be drawn in a vertical plane through the unconformity as shown in Figure 7.13, using the methods already established, at right angles to the strike of both sequences.

Such a cross-section may also provide an adequate representation of an unconformity even if the two sequences differ in trend. Thus, if most of the folding and faulting had occurred before the deposition of the rocks lying above the unconformity, so that now these rocks have only a low dip, it would be possible to draw a vertical cross-section at right angles to the structures developed in the underlying rocks, without introducing too much error. Likewise, if most of the folding and faulting had occurred after the deposition of the rocks lying above the unconformity, so that there is only a slight discordance between the two sequences, it would be possible to draw a vertical cross-section at right angles to the structures developed in the overlying rocks, without introducing too much error. A compromise must be made in choosing the line for such a cross-section. Ideally, it should pass across as many formations as possible in the rocks lying below the unconformity, while maintaining such a course that the structure of the overlying rocks can also be shown without too much difficulty.

Once a suitable line has been chosen, the structure of the rocks exposed on either side of the unconformity is entered on the cross-section. This can be done in a straightforward fashion wherever these rocks are encountered along the line of section. The unconformity at the base of the overlying sequence should then be projected up its dip across the outcrop of the older rocks, above the present level of erosion. The contacts between the various formations lying above the unconformity can be extended in a similar manner. Any overlap associated with the unconformity should be taken into account in the construction of this part of the cross-section. The structure of the rocks lying below the unconformity should next be projected upwards above the present topography, to be intercepted by the unconformity at the base of the overlying sequence. Finally, the cross-section can be completed by drawing in the structure of the rocks lying below the unconformity, where they are found beneath the overlying rocks.

A subjective element is obviously introduced into the construction of such a cross-section. Thus, it is necessary to decide whether or not the folds and faults developed in the rocks below the unconformity also extend upwards to affect the overlying rocks, away from their outcrop. This depends on when these structures were formed in relation to a particular unconformity.

Pre-unconformity Folding and Faulting.

If most of the folding and faulting occurred prior to the deposition of the rocks lying above the unconformity, there would be a marked difference in structural style between the two sequences, as shown by the outcrop pattern. It is then found that the rocks below the unconformity are more disturbed by folding and faulting than the overlying rocks, as shown in Figure 7.13. Such a difference in structural style would also result in a marked discordance across the unconformity, as expressed by the overstep shown by the unconformity as it is traced along its outcrop, away from the cross-section. Such evidence can be used to reconstruct the nature of the pre-unconformity structures which are developed in the underlying rocks. These structures may then be projected parallel to their plunge so that they can be shown on the cross-section, truncated by the rocks lying above the unconformity.

This may be done by tracing the subsurface outcrop of the various formations below the unconformity, until they intersect the cross-section. Such a procedure would obviously be facilitated by the presence of inliers, which would allow these structures to be traced for a greater distance than would otherwise be possible. Ideally, the line of section should be chosen to pass through such inliers. It may also be possible to trace the structures developed in the underlying rocks away from the unconformity, so that they can be shown along strike on the cross-section. The unconformity can then be projected up-dip across the outcrop of the underlying rocks in such a way that it cuts across these structures as they are projected upwards from the present level of erosion. It is usually assumed that the overlying rocks do not change in dip as they are traced in this direction, unless there is any evidence to the contrary. Such evidence might be provided by the presence of outliers, which could show the rocks above the unconformity to be fold-ed or faulted. Ideally, the line of section should also be chosen to pass through these outliers. It may be possible to project any folds which are developed in the overlying rocks, so that they can be shown up-plunge on the cross-section, cutting across the structures developed in the underlying rocks.

It should be realised that the angular discordance marking an uncon-formity develops as a result of tilting or folding rather than faulting. Accordingly, an unconformity may show only a slight discordance between the two sequences, even although it cuts across a major pre-unconformity fault. The presence of such a fault can be recognised wherever the uncon-formity oversteps abruptly across two formations, which are not in stratigraphic order, as it is traced across a particular fault-line.

Post-unconformity Folding and Faulting.

If most of the folding and faulting occurred after the deposition of the rocks lying above the unconformity, there would be only a slight difference in structural style between the two sequences. This would be reflected in the outcrop pattern, which would show the folding and faulting to affect the rocks of both sequences as a single unit. This means that only a slight discordance would be developed between the two sequences, marking the unconformity, as the result of pre-unconformity tilting. The folds and faults would then be found to affect the unconformity and the overlying rocks as they are traced across the geological map as a single set of struc-tures. The only difficulty likely to be encountered in the construction of a cross-section in such a case concerns the angular discordance developed be-tween the two sequences. However, this can be determined from the nature of the overstep associated with the unconformity, as observed along the cross-section itself.

Synoptic Cross-sections and their Palinspastic Reconstruction

Although a structural cross-section can be constructed to illustrate the nature of the structures that developed in response to a certain history of geological events, it is often difficult to choose a satisfactory line which would show all the salient features of the geology. However, there is an alternative approach to this problem. This arises from the fact that the geological structure of any area is defined by a number of stratigraphic se-quences, which have certain stratigraphic and structural relationships with one another. These relationships are not always encountered along a par-ticular line, nor can they always be projected to appear on the cross-section. However, it is usually possible to construct a schematic cross-section, show-ing how the various stratigraphic sequences are related to one another. This can be termed a synoptic cross-section, since it is intended to provide a synopsis of the stratigraphic and structural relationships which exist between the various sedimentary sequences within a particular area.

Construction of a Synoptic Cross-section.

It has already been shown how an outcrop pattern as shown in Figure 7.23 can be analysed in terms of a number of stratigraphic sequences, separated from one another by unconformities, which together form the stratigraphic column for the area under consideration. The stratigraphic and structural relationships between these sequences are then defined according to how the unconformity at the base of each sequence cuts across the geological structures developed in the underlying rocks. These unconformities can be used as stratigraphic markers to date the folding, faulting, igneous intrusion and regional metamorphism, which may affect these stratigraphic sequences. The synoptic cross-section is intended to show the geological structure which develops in response to the history of geological events that can be erected according to all this evidence. In particular, it would show the nature of the overstep which is developed at the base of each stratigraphic sequence, since it is this overstep which defines how folding and faulting affected the underlying rocks, prior to the deposition of the overlying sequence.

Ideally, a synoptic cross-section should be drawn at right angles to the subsurface outcrops which are developed by the underlying rocks against an unconformity. This usually means that the cross-section is constructed at right angles to the trend of the geological structures which occur below the unconformity. Obviously, if more than one unconformity is present, a compromise may need to be made. Under these circumstances, most weight should be given to the unconformity which shows the greatest degree of structural discordance. This might be represented by the present topography. The cross-section should be drawn to show the true nature of the overstep associated with this particular unconformity. The structural features associated with the other unconformities can then be projected so that they would be shown at an oblique angle on the cross-section. Any distortion arising from this manner of projection should be corrected.

Although the general orientation of the cross-section needs to be decided, this does not imply that it must follow a certain line across the geological map in the manner of a structural cross-section, since this might defeat the whole object of the exercise. Moreover, the cross-section does not need to be drawn true to scale, since it is only intended to illustrate the stratigraphic and structural relationships between the various sequences in the form of a schematic diagram. It can therefore be drawn in a stylised manner, without showing too much detail, as shown in Figure 7.24.

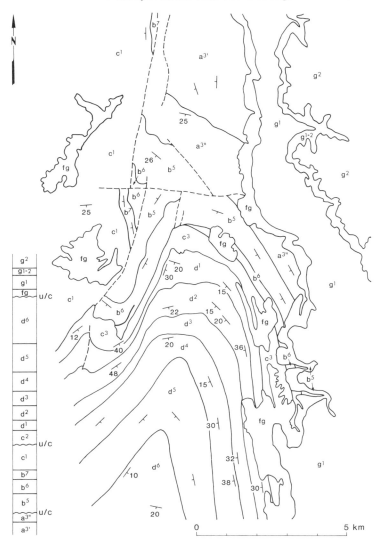

FIG. 7.23. Geological map and stratigraphic column illustrating the outcrop pattern associated with development of two major unconformities. (Redrawn with modification from Sheet 251 of the Geological Survey of England and Wales.)

Palinspastic Reconstruction of the Geological Structure.

Once a synoptic cross-section has been constructed, it can be used to show how the geological structure developed in response to a certain history of geological events. Thus, it may be assumed as usual that each stratigraphic sequence was originally deposited with little or no dip, unless there is any evidence to the contrary. The effect of subsequent earth

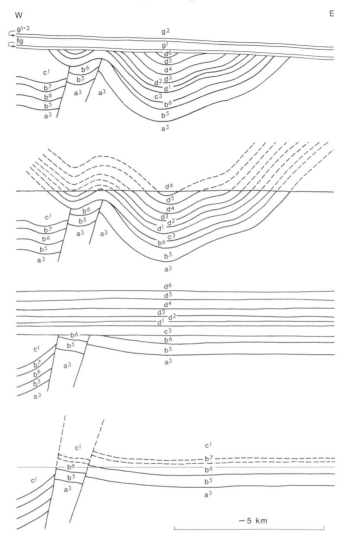

Fig. 7.24. *Upper diagram:* synoptic cross-section showing the stratigraphic and structural features of the area shown in Figure 7.23. *Lower diagrams:* palinspastic reconstructions of the synoptic cross-section showing how the geological structure evolved in response to two major episodes of earth movements.

movements can then be removed from the geological record by restoring the bedding of each stratigraphic sequence in its turn to a horizontal position in such a way that the rocks regain their original continuity. It should be borne in mind that it is difficult to do this objectively in the case of faulting. A similar adjustment can be made at the same time to the geological structure of the underlying rocks. By removing the effects of any later episodes of folding and faulting in this way, the underlying rocks would regain the geological structure that they had prior to the deposition of the overlying sequence of sedimentary rocks. The nature of the pre-unconformity structure can then be determined. This can be attempted for each stratigraphic sequence in its turn, working backwards from the present structure as shown by the synoptic cross-section.

The result would be a series of palinspastic reconstructions showing the geological structure as it existed at certain times during the geological history of the area (see Figure 7.24). Each reconstruction should be drawn so that a particular sequence of sedimentary rocks rests in a horizontal position on top of an unconformity, cutting across the geological structures which are developed in the underlying rocks. The form of such a reconstruction serves to emphasise the episodic nature of any geological history. This is defined by repeated episodes of uplift and erosion, accompanied by folding and faulting of the underlying rocks, which are followed in each case by subsidence and the further deposition of sedimentary rocks, to form a stratigraphic sequence resting unconformably on the rocks so affected. A palinspastic reconstruction can be attempted for each gap in the stratigraphic record, as marked by the development of an angular unconformity. The whole series would then illustrate the structural evolution of the area in response to such a geological history.

Nature of Earth Movements

Such an interpretation of geological structure in terms of its historical evolution properly forms the foundation of an all-embracing branch of geology known as tectonics. This considers how the geological structure of the earth's crust evolves in response to such processes as uplift and erosion, subsidence and deposition, tilting and folding, deformation and metamorphism, and igneous activity, together with continental drift and its consequences. As appropriate to its derivation from the Greek word for a builder, the study of tectonics is concerned with the architecture of the earth's crust and how it arose. So defined, tectonics covers a much wider canvas than the mere study of structural geology. It is particularly concerned with the effect of earth movements on the geological evolution of the earth's crust, while it also considers the structural setting of individual regions where these movements have taken place. Tectonics therefore uses the evidence of the geological record to establish what earth movements have occurred during the course of geological history, in an attempt to understand what internal processes were responsible for these movements in the first place.

Earth movements can be divided into two broad categories, known respectively as epeirogenesis and orogenesis. Epeirogenesis (or epeirogeny) is the name given to vertical movements of uplift and subsidence, which affect wide areas of the continental crust. Although such movements continue to take place throughout the course of geological history, their character changes in that uplift repeatedly alternates with subsidence in any one region, while uplift and subsidence affect different regions at any one time. The nature of these movements can be determined from an examination of the stratigraphic record.

It has just been shown how this consists of a number of sedimentary sequences, separated from one another by angular unconformities, at least in structural terms. The interpretation placed on such an unconformity provides us with evidence of uplift from the geological record. Likewise, the deposition of sedimentary rocks can only take place where subsidence allows sufficient space for their accumulation. The presence of stratigraphic sequences, separated from one another by angular unconformities, can therefore be taken as evidence in the geological record of subsidence and deposition, interrupted by periods of uplift and erosion. Moreover, it is the uplift of pre-existing rocks which provides a source for the sedimentary rocks which tend to accumulate as a result of subsidence. This means in effect that deposition occurs in sedimentary basins, formed as a result of differential subsidence, that are separated from one another by intervening uplifts. Although such movements of differential uplift and subsidence are epeirogenic in character, they can lead to the development of structural discordances on a regional scale wherever they have caused the tilting and warping of sedimentary rocks lying below an unconformity, prior to the deposition of the overlying rocks. This means that such an unconformity, and other breaks in the stratigraphic record, may also be taken as evidence for epeirogenic movements in the geological record.

Epeirogenesis can be compared with orogenesis, which is defined as mountain building according to its etymological derivation. However, orogenesis is a term which was coined before a clear distinction was drawn between the differential uplift which produces a mountain range and the folding and faulting which affect its constituent rocks. Since the uplift of a mountain range may reasonably be regarded as the result of epeirogenic movements, it is now accepted that orogenesis should refer in particular to the folding and thrusting which so alter the internal structure of the rocks forming a mountain range. Such movements are therefore considered to be orogenic in character. They are often accompanied by regional metamorphism and igneous intrusion.

The evidence of orogenic movements may be found in the geological record wherever intensely folded and faulted rocks, often affected by igneous intrusion and regional metamorphism, are overlain unconformably by less disturbed sedimentary rocks. This means in effect that it is the degree of structural discordance existing across an angular unconformity which allows a distinction to be drawn between epeirogenic and orogenic movements. Although epeirogenic movements affect all but the most stable parts of a continent, it is invariably found that orogenic movements are much more restricted in their effects. The sedimentary rocks deposited during the course of a particular episode of geological history are therefore to be affected by varying amounts of folding and faulting, if they are traced over a wide enough area. This allows cratons to be distinguished from orogenic belts.

Cratons and Orogenic Belts.

Cratons are defined as the relatively stable segments of the continental crust, which have not been affected to any extent by orogenic movements. They can therefore be distinguished from orogenic belts of the same age, which are so affected by folding and faulting. The sedimentary rocks deposited within the craton form a stratigraphic sequence which usually retains a flat-lying attitude, since it is only affected to a slight degree by folding and faulting. Such sequences tend to be relatively thin, rarely exceeding a thickness of more than a few kilometres at the very most.

The flat-lying sediments deposited on the craton are generally underlain by older rocks, from which they are separated by an angular unconformity. The underlying rocks form a basement to the craton, so that they are known as basement rocks. They have usually been affected to a considerable extent by folding, faulting, deformation, regional metamorphism and igneous intrusion, prior to the deposition of the overlying sediments. The latter form a cover to the basement rocks, so that they are known as cover rocks. The basement rocks are only exposed at the earth's surface where the cover rocks have been eroded away, assuming that they were deposited there in the first place. Such areas are called shields wherever the basement rocks are Precambrian in age. The name implies that the basement rocks have a broad arch-like disposition, similar in form to a medieval shield.

Orogenic belts can be distinguished wherever intense folding and thrusting have affected a sedimentary sequence after its deposition. This means that an orogenic belt can simply be defined as an area where the rocks have been affected by orogenic movements. It is commonly found that orogenic belts occur around the margins of the continental cratons, separating continents from one another or juxtaposed between continent and ocean. They are usually rather long and narrow, since they can often be traced for several thousand kilometres while they are rarely more than a thousand kilometres in width. Orogenic belts commonly maintain a common trend throughout their length, even although their margins are often formed by a series of convex arcs, facing towards the craton.

The sedimentary sequence affected by folding and thrusting within an orogenic belt is commonly much thicker than the corresponding sequence which is found on the adjacent craton. These rocks are generally affected by varying degrees of deformation, regional metamorphism and igneous intrusion. The deformation and regional metamorphism may so alter any basement rocks, which are found below the sedimentary cover-rocks, that their original structure is partially obscured or entirely lost. This process is known as reworking or reactivation. It results in the basement rocks taking on a structure similar to the cover rocks. This means that the sedimentary rocks no longer form a true cover to the basement rocks, so that they are then best distinguished as supra-crustal rocks.

CHAPTER 8

Cratons and Orogenic Belts

Nature of Tectonic Maps

THE structural features shown by cratons and orogenic belts can be illustrated by the construction of a tectonic map. This is commonly drawn on a relatively small scale, so that it may cover as large an area as a continent. Such a map differs in several respects from a geological map. In particular, it does not show the outcrop of lithostratigraphic units, except on a very large scale, since it is intended to depict the tectonic character and geological structure of the rocks forming these units. It should be emphasised that a systematic methodology has not yet been estabished for the construction of tectonic maps, in comparison with the rather standard procedures which are now used in geological mapping of a more conventional kind.

The features shown by a tectonic map occur on different scales. The primary features can be considered to arise from the tectonic division of the continental crust into cratons and orogenic belts. Such a division is usually shown by differences in colouring between these areas. The depiction of cratonic areas is commonly intended to show simply the structure developed by the upper surface of the underlying basement rocks. This may be done using structure contours drawn on this surface, with the added possibility that the areas lying between these contours can be coloured with layer tints to indicate the varying depth to the basement. Alternatively, colours may be used to distinguish the different sequences of sedimentary formations that are present on the craton, separated from one another by angular unconformities of regional extent. The outcrop of basement rocks could also be coloured according to their age.

The complex features shown by orogenic belts are more difficult to depict on a tectonic map. They are often coloured to show the age of the orogenic movements which marked the final climax in their development. This means in effect that each orogenic belt is identified by the use of a particular colour. Different shades of this colour are commonly used to distinguish volcanic terrains within each fold belt, while areas of regional metamorphism and migmatisation may also be shown separately. Any basement rocks occurring within such an orogenic belt can also be coloured according to their relative age. Some maps may show what are known as structural stages, corresponding to different parts of the sedimentary sequence within an orogenic belt, even although they may not be separated from one another by angular unconformities. Finally, the nature of the intrusive rocks occurring within a particular belt can be distinguished by different colours.

The secondary features shown by a tectonic map are the individual structures which occur as a response to folding, faulting and the like. The depiction of these structures is relatively straight forward, since it can be done using the standard methods employed in geological mapping. These structures are generally shown in some detail, depending on the scale of the map itself, so allowing a clear picture of the regional structure to be determined from the study of a tectonic map. Difficulties are only likely to be encountered in the internal parts of orogenic belts, where the geological structure is extremely complex.

Structural Features of the Craton

Since the sedimentary rocks of cratonic areas exhibit little in the way of folding or faulting, their geological structure develops mainly in response to differential uplift and subsidence. As already described, these processes lead to the development of sedimentary basins, separated from one another by areas of relative uplift. The structure of these basins is usually shown on a tectonic map by means of structure contours, drawn on top of the basement rocks. These contours give the original form of the sedimentary basin and its surroundings, as modified by the effects of any folding or faulting which has occurred after deposition came to an end. Whether or not the presence of such a basin can be inferred from the outcrop pattern shown by a geological map depends on the present level of erosion.

If uplift and erosion has not affected a sedimentary basin to any extent, its outlines tend to be hidden by a thin veneer of sedimentary rocks, which were deposited as the youngest formations within such a basin. By way of contrast, if uplift and erosion have removed the youngest rocks deposited within a sedimentary basin, its outlines are exposed to view. A geological map then shows the central parts of the sedimentary basin to be marked by the outcrop of the youngest formations, while the surrounding uplifts correspond to the outcrop of the oldest formations, within a particular region. Such an outcrop pattern arises naturally from the structure of a sedimentary basin, which develops in response to differential subsidence. However, this structure may be accentuated by any folding or faulting which occurs after the sedimentary rocks were deposited within the basin. Since it is the older basins which are most likely to be affected by uplift and erosion it is these basins which are most commonly exposed to view in this way.

Folding and Faulting.

Although a general lack of folds and faults is a characteristic feature of the craton, this does not preclude the local development of such structures on a modest scale. Folding generally occurs in an irregular manner, as shown by the structure contours drawn on a particular horizon within the stratigraphic sequence. It may take the form of irregular domes and basins, or elongate anticlines and synclines may be formed with irregular and variable trends. The dips on the flanks of these structures are unlikely to exceed a few degrees at the very most.

Folds may be formed by differential compaction wherever the underlying rocks of the basement show a certain amount of topographic relief. The sedimentary rocks are deposited so that their thickness is reduced over any culminations in the underlying topography. The sedimentary sequence is then affected by differing amounts of compaction, according to its thickness at any one place, as shown in Figure 8.1. This means that most compaction occurs away from the culminations shown by the underlying rocks, where the sedimentary sequence has its greatest thickness. A stratigraphic horizon would be distorted so that it reflected the topographic relief developed by the basement rocks. This causes anticlines to form as compaction folds over any buried hills in the underlying basement. The form of such folds is often rather irregular, so that they commonly occur as domes. They are only developed at a low level in the cover rocks, since they tend to lose their form as they are traced upwards into higher levels of the sedimentary sequence. This is accomplished by the sedimentary layers thinning over the anticlinal areas, so that the structure has the form of a Class 1A fold. It is a common feature of all these folds that the dips developed in response to differential compaction are always less than the slope of the underlying basement.

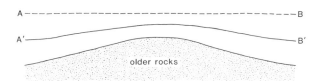

FIG. 8.1. Development of compaction folds. AB: Original surface of deposition. A'B': Surface of deposition as represented by the corresponding bedding plane after differential compaction of underlying sediments.

Comparable structures are developed wherever there are differences in the thickness of a stratigraphic sequence, which allows differential compaction to take place. Such differences in stratigraphic thickness may simply arise from differential subsidence during the development of a sedimentary basin, or they may be caused by any penecontemporaneous faulting which occurs during the deposition of the sedimentary rocks within such a basin. The form of any structures developed in response to these movements would be accentuated by the differential compaction which occurs as a result of such differences in stratigraphic thickness. Differential compaction may also occur wherever there are lenticular bodies of conglomerate, sandstone or limestone, interbedded within a sedimentary sequence, which undergo less compaction than the surrounding material.

Faulting tends to occur in a more regular manner, as the result of normal dip-slip movements. This commonly leads to the development of fault zones, in which all the individual faults have approximately the same trend. The sedimentary rocks lying within these zones are usually affected by block faulting, so that horsts and graben are produced as a result. Alternatively, a single down-faulted block may be formed by rift faulting, flanked on either side by step faults. There is a gradual transition between these two extremes in nature. These fault zones often appear to cut across the craton in a random fashion, so that they are not related in any way to the form of the sedimentary basins which define its structure. However, the present-day rift valleys form an exception to this statement, since they are consistently developed along uplifts which separate sedimentary basins of a fairly recent age from one another.

Although faults may be developed in the rocks of the underlying basement, it is commonly found that they are replaced by monoclinal folds as they are traced upwards into the sedimentary cover-rocks. Such structures are more likely to be associated with the development of upthrusts rather than normal faults. Alternatively, if the faulting is strike-slip rather than dip-slip, en echelon zones of periclinal folds may be developed as brachyanticlines in the cover rocks.

Salt Domes

The craton may also be marked by the presence of salt domes, wherever the stratigraphic sequence contains deposits of rock salt and other evaporites at depth. These structures are each formed by a columnar body of rock salt, penetrating upwards into the overlying rocks in the form of a vertical plug, as shown in Figure 8.2. This usually has a diameter of 2-3 kilometres, although some examples are much larger. Each plug may extend downwards to a depth of several kilometres, corresponding to the stratigraphic level of the rock salt which acts as its source. The walls of most salt domes dip outwards at a steep angle so that they become wider at depth. However, occasional examples appear to be spindle-shaped in that they have become completely detached from their source at depth. The upper terminations of most salt domes are flat-topped or dome-shaped. Commonly, an overhang is developed as a salt dome takes on a bulbous form

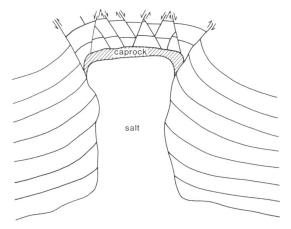

Fig. 8.2. Structural features shown by a salt dome.

near the surface. The tops of many salt domes are marked by a sheath of cap rock, which is formed by the solution of rock salt above the level of the ground water, leaving an insoluble residue of limestone, anhydrite and gypsum.

The sedimentary rocks surrounding a salt dome show all the effects of its upward penetration as an intrusive or diapiric body, which cuts across the structures developed in the country rocks. In particular, sedimentary formations become truncated against the walls of a salt dome, while they are dragged upwards into a steeper attitude in its vicinity. This may be accompanied by the development of a rim syncline as an arcuate structure encircling the salt dome at a slightly greater distance. Such a structure is thought to be formed by the withdrawal of rock salt from the source layer around the salt dome as it moves upwards into the overlying rocks. A complex pattern of normal dip-slip faults is often developed in the sedimentary rocks overlying a salt dome. These rocks are usually arched over the salt dome, so that their structure is defined by a dome in the overlying rocks. Tangential and radial faults are commonly found as an expression of the horizontal extension which affects these rocks as the salt dome moves towards the surface.

Salt domes are usually distributed in a haphazard manner wherever they penetrate flat-lying sequences of sedimentary rocks. They may be numbered in hundreds if a wide enough area is considered. However, similar structures are also developed in sedimentary sequences which have been affected by folding and faulting as a result of horizontal compression, along the margins of orogenic belts. Under these circumstances, the individual salt domes take on an elongate form, so that they might better be termed salt anticlines (see Fig. 8.3). The parallel alignment of these bodies defines a regional trend at right angles to the inferred direction of horizontal compression. This trend is shared by the folds and faults which are otherwise developed in these rocks.

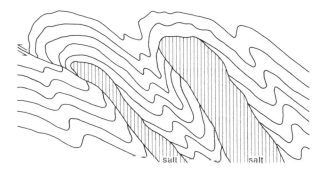

FIG. 8.3. Diapiric intrusions of an underlying salt horizon in an area affected by folding and overthrust faulting.

Salt domes arise from the diference in density between rock salt and the overlying rocks, which develops once these rocks have undergone a certain amount of compaction. Such a difference in density becomes important once sedimentary rocks have been deposited on top of the layer of rock salt to a thickness of rather less than a kilometre. Further compaction under an increasing load of sedimentary rocks causes a gravitational instability to develop, since the rock salt is less dense than the overlying rocks. The buoyancy of the rock salt then allows slight irregularities in the upper surface of the source layer to develop eventually into salt domes, by the upward flow of rock salt into areas of relatively low pressure. This process is aided by the relative ease with which rock salt can flow as a solid under differential pressure.

The irregularities in the upper surface of the source layer may be random in their distribution, so that the salt domes form with a haphazard distribution, as commonly found. However, they may develop in response to tectonic forces, so that the salt domes tend to form elongate bodies with a parallel alignment. This may be the case even if folding or faulting has not otherwise affected the sedimentary rocks to any extent. If folds do develop, the rock salt acts as a highly incompetent material, which flows in order to accommodate the folding of the competent layers. This is commonly accompanied by the upward intrusion of the rock salt along thrust planes, forming cores of injected material which cut across the bedding of the overlying formations. These structures are known as piercement folds.

Block Faulting and Folding

Although the central parts of a craton may only be affected slightly by epeirogenic movements, it is commonly found that its margins adjacent to an orogenic belt undergo a certain amount of folding and faulting, accompanied by epeirogenic movements of a more intense nature. It should be emphasised that the structural features, which develop as a result of these movements, resemble those shown by the rocks of the craton itself, even although their form is much accentuated. These marginal regions should therefore still be regarded as part of the craton. The tectonics are often termed Germanic (or Germanotype), in comparison with the Alpine (or Alpinotype) tectonics characteristics of orogenic belts.

The marginal areas of the craton are commonly affected by epeirogenic movements which produce an irregular mosaic of sedimentary basins, separated from one another by uplifts. The sedimentary basins are often rather narrow, while they may be affected by a considerable amount of subsidence. The uplifts are usually positive areas undergoing erosion, which act as a local source for the detrital sediments deposited in the nearby basins. The outlines of these basins and uplifts may be controlled to some extent by structural trends in the underlying basement rocks. It is commonly the case that volcanic activity accompanies the deposition of sedimentary rocks. This would produce lava flows and pyroclastic deposits interbedded with the sedimentary rocks as part of the stratigraphic sequence. Volcanic necks and minor intrusions are likely to be associated with the extrusive rocks.

The structural features displayed by the sedimentary cover rocks are mainly the result of block faulting, which affects the underlying basement. The faulting tends to follow the structural trends which are developed in the basement rocks, so that the fault pattern is often rather complex. More than one trend may be apparent so that an irregular pattern is formed by upfaulted horsts of basement rocks, separated from one another by downfaulted basins of sedimentary rocks. Whether or not the basement rocks are exposed as a result of this faultig depends on the present level of erosion.

It is uncertain whether the margins of the up-faulted blocks of basement rocks are formed by normal or reverse dip-slip faults although upthrusting rather than normal faulting often appears to be the dominant mechanism. Although these faults would dip steeply under the basement rocks at depth, they may be curved so that they become flat-lying as they are traced towards the surface. Even if this is overall an original feature, the curved form of these faults may be accentuated by the lateral spreading of the hanging wall, which is formed by the basement rocks at a high structural level, under the influence of gravity. Normal dip-slip faulting may also occur within the upfaulted block, extending upwards to affect the sedimentary rocks of the cover. This can simply be due to an arching of the upfaulted block, while the effects of lateral spreading may also be important

The flat-lying rocks forming the sedimentary cover of these horsts are usually only affected by normal dip-slip faulting. However, the sedimentary rocks become increasingly disturbed as they are traced towards the margins of the up-faulted blocks. These are commonly marked by the development of monoclinal folds in the sedimentary rocks, while the faults forming the margins of horsts at depth extend upwards to affect the overlying cover rocks nearer the surface. These faults may die out as they are traced upwards into the sedimentary cover, leaving only a monoclinal fold to mark their presence.

The sedimentary cover rocks are usually preserved in the structural basins which are developed between the up-faulted blocks of the basement. The folds affecting the sedimentary rocks often disappear as they are traced away from the margins of such a basin, leaving an area of flat-lying sedimentary rocks in its centre. However, this only happens if the structural basin is relatively broad. The sedimentary rocks confined within a narrower basin are commonly found to be tightly folded. There is usually a complete transition between the two extremes, so that most structural basins show a considerable amount of folding away from their margins. The folding is commonly periclinal in nature, so that elongate domes and basins are developed with irregular and variable trends. By way of contrast, the folding along the margins of the structural basin is often more regular in its development. This would produce anticlines and synclines, trending parallel to its margins. Such structures may be formed as a result of lateral compression, which occurs locally in response to the upthrusting of the basement rocks.

However, it is likely that the complex and intricate pattern of folding and faulting which often affects the sedimentary rocks of such regions may occur as the horsts of basement rocks move in relation to one another under the influence of regional forces. This effect is often regarded as analogous to the jostling of ice floes, as they are driven together by the wind. However, it is a moot point if such an analogy can also account for the upthrusting which appears to be a characteristic feature of such regions of block faulting. Other mechanisms may also be responsible. Recent geophysical surveys have shown, for example, that some basement uplifts in block-faulted regions are underlain by shallowly dipping faults to depths of 20-25 km. This implies that the reverse faults, seen to dip at a low angle near the surface, do not steepen at depth as previously thought, at least in some areas. Such block-faulting is presumably the surface expression of overthrusting at deep levels within the crust.

Marginal Zones of Orogenic Belts

The central parts of an orogenic belt are commonly marked by the outcrop of intensely deformed and highly folded rocks, which have also been affected to a considerable extent by regional metamorphism and plutonic activity. These areas are flanked by the external zones of the orogenic belt, forming its margins against the craton. The rocks lying within these zones are typically affected by folding and thrust-faulting, but they are not intensely deformed, as described in Chapter 5. There is a general lack of regional metamorphism and plutonic activity. It may be inferred that the original structure of the basement rocks, which underlie the folded and faulted cover rocks within these regions, is not greatly altered by the orogenic movements.

Some belts show a rapid decline in intensity of the orogenic movements towards the craton. The external zones of these belts would be relatively narrow. They are often marked by a narrow zone of thrust faulting, across which the folding and regional metamorphism affecting the orogenic belt dies out. The external zones of other belts are much wider. These zones are marked by wide-spread folding and thrusting, which affects an extensive region along the margins of the orogenic belt against the craton. It is often difficult to judge where the boundary should be drawn between such an external zone and the adjacent craton, since the structural features characteristic of the former area may be found to die out very gradually as they are traced away from the orogenic belt.

It has already been remarked that folding and thrust-faulting are a characterisitc feature of the external zones of an orogenic belt. This is usually accompanied by décollement, so that the sedimentary cover rocks are detached from the underlying basement along a horizon of incompetent material. This means that the basement rocks are not affected by the folding and thrusting. The structural features developed by the overlying rocks may be formed in response to folding or thrusting, or a combination of both processes acting together. It is commonly found that folding is replaced by thrusting as the rocks forming the external zone of an orogenic belt are traced along their strike.

Such zones of folding and thrusting along the margins of an orogenic belt are rarely more than 250 kilometres in width at the very most. The structural trends developed within these elongate zones of folding and thrusting are parallel to the outlines of the orogenic belt itself. This means in effect that the folds and thrust-faults have a rectinlinear arrangement wherever the orogenic belt has a straight boundary, while they would form an arcuate pattern wherever its margin is formed by a series of convex arcs, facing towards the craton. Individual structures can rarely be traced for a long distance parallel to the regional trend. Instead, as they die out, individual folds and thrust-faults tend to be replaced by other structures. The folding tends to produce a series of elongate periclines in the form of doubly-plunging brachy-anticlines and brachy-synclines, so that one fold is replaced by another along the strike. Folds may also be replaced by thrust faults. Strike-slip faults are often found as oblique structures within thrust zones and fold belts, lying at a high angle to the regional trend. These faults can also take up the folding and thrust-faulting if they develop at the same time.

Nature of the Folding and Thrusting.

The folding tends to occur as a result of flexural slip, affecting the competent layers within the sedimentary cover rocks, to produce parallel folds belonging to Class 1B. The structural features shown by these folds have already been described as arising from their disharmonic nature. However, it may be noted that the style of these folds depends to a considerable extent on the nature of the sedimentary sequence. If it is underlain by a particularly incompetent horizon, the folding of the whole sequence tends to produce anticlinal box-folds, separated from one another by broad synclines, which become narrower as the folding becomes more intense towards the orogenic belt itself. Such structures are developed in the Jura Mountains. Alternatively, if there are incompetent horizons separating the competent layers from one another within the stratigraphic sequence, the folding tends to produce concentric folds with rounded hinge zones and relatively short fold limbs. These folds would become tighter as they were traced towards the orogenic belt. They are usually inclined so that their axial planes dip towards the orogenic belt.

The folding of the sedimentary cover would be accommodated in either case by décollement along a suitable horizon at the base of the stratigraphic sequence. This would form a sole thrust on which the displacement increases as it is traced towards the orogenic belt. Similar bedding-plane thrusts are likely to be formed at a higher level wherever there are incompetent horizons developed within the stratigraphic sequence. It is commonly found that the bedding-plane thrusts may cut across the competent layers to form a series of step thrusts, ascending the stratigraphic sequence towards the foreland. Folds are likely to be generated in the overlying rocks as they move over such a step thrust. It is commonly the case that bedding-plane thrusts end in a splay of high-angle reverse faults, forming an imbricate zone against the undeformed rocks of the foreland. Splay faulting of this type can extend throughout the thrust zone, so that a series of concave-upward thrusts is formed above the sole. It is these thrusts which tend to be folded at a later stage, when the forward movements of the thrust sheets becomes impeded as a result of the thrusting which has already occurred.

Effects on the Basement.

Structures of a different kind are produced wherever there is no suitable horizon for décollement to occur on a sole thrust near the base of the stratigraphic sequence. Under these circumstances, the basement rocks may well be affected by the deformation, as it dies out gradually away from the orogenic belt. It is quite likely that the basement rocks respond to this deformation by faulting rather than folding. This produces a series of fault blocks, controlling the form of the folds that would otherwise develop freely in the cover rocks. The tectonic style may therefore resemble that produced by block faulting, even although the folding is likely to be more intense in the present case. The lack of any suitable horizon for décollement would favour the development of chevron folds, which can maintain their profile at depth, even although they are formed by flexural slip. This factor can also account for the common development of low-angle reverse faults in such fold belts. These thrusts are often formed at an early stage in the deformation, so that they become folded at a later stage.

Slate Belts

The central parts of an orogenic belt differ from the marginal zones of folding and thrusting in that the geological structure tends to be much more complex. There are several reasons why this should be so. Firstly, although the folding and thrusting at the margins of an orogenic belt are often only developed in response to a single phase of movement, it is commonly found that the internal zones are affected by a structural history of more complexity. In particular, folding may be followed by refolding, so that several generations of folds are developed in response to a polyphase history of deformation. Secondly, the internal zones of an orogenic belt are typically affected by regional metamorphism and migmatisation. This marks an influx of thermal energy into the crust, which allows the rocks to deform more easily under higher temperatures. The deformation is associated with the development of secondary structures such as cleavages under these conditions. Thirdly, it is commonly found that the central parts of an orogenic belt are marked by the forceful intrusion of granitic batholiths, which can modify the geological structure of the country rocks to a considerable extent as described in Chapter 6. Finally, there is also a complex interaction between the supracrustal cover rocks and the underlying basement rocks, which is generally lacking in the marginal zones of folding and thrusting.

Unless the rocks forming the central parts of an orogenic belt have been thrust over the marginal zones of folding and thrust-faulting, there is often a gradual transition between these divisions of an orogenic belt. This transition is typically marked by the gradual development of a slaty cleavage, often starting in the guise of a fracture cleavage, which forms in response to deformation under metamorphic conditions of the lowest grade. As the name suggests, the former type of cleavage is characteristic of slaty rocks which have a marked tendency to split into very thin slabs along a whole series of parallel planes. Since these slabs are commonly used as roofing slates, any rock with such a fissility is generally known as a slate. It is usually produced from a fine-grained rock which would otherwise be a shale or a volcanic ash. The presence of such a slaty cleavage in the rocks of a particular area serves to define what is known as a slate belt.

Nature of a Slaty Cleavage.

The fissile nature of a slate depends on the presence of an internal fabric, which is defined by dimensional orientation of inequant grains. It is this fabric that forms the structure known as a slaty cleavage. Typically, the fabric of most slates is defined by the dimensional orientation of layer-silicate minerals such as sericite, muscovite, chlorite and biotite, as shown in Figure 8.4A. These minerals generally occur as flakes with their longer dimensions parallel to (001). The slaty cleavage is then defined by a common orientation of the (001) planes throughout the rock in such a way that all these planes tend to be parallel to one another. These minerals may be uniformly distributed, or they may be concentrated in discrete zones to form mica films which are parallel to the slaty cleavage. These mica films often form a braided network of anastomosing planes on a fine scale, which is only statistically parallel to the slaty cleavage. Such a structure is known as a spaced cleavage. It can be compared with fracture cleavage which is formed by a series of closely spaced fractures in the rock, lacking any dimensional orientation of the constituent minerals.

It is commonly found that detrital grains of quartz and felspar are so deformed in slaty rocks that they become flattened in the plane of the slaty cleavage. Such grains often have lenticular outlines, while they may be separated from one another by discrete mica films. The dimensional orientation of inequant grains of quartz, felspar and carbonate also serves to define a planar structure similar to slaty cleavage, which is often developed in mica-poor rocks such as quartzites and metamorphic limestone.

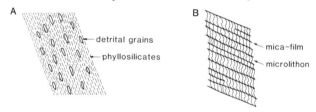

FIG. 8.4. Textural features seen in thin-section characteristic of (A) slaty cleavage and (B) crenulation cleavage.

Development of the Stretching Direction.

The deformation accompanying the development of a slaty cleavage can be measured, wherever it is possible to determine the original shape of deformed objects in the rock. Such objects are known as strain markers. Spherical objects such as oolites and reduction spots make good strain markers, while fossils and other objects may also be used. The study of spherical objects in particular shows that a slaty cleavage commonly develops at right angles to the direction of maximum compression in the rock, as shown in Figure 8.5. This means that the direction of maximum

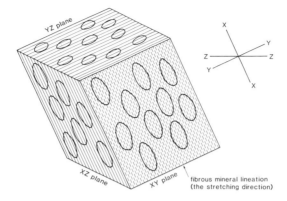

FIG. 8.5. Strain markers showing that slaty cleavage typically develops at right angles to the direction of maximum compression (Z) in the rock. Note that a fibrous mineral lineation (the stretching direction) defines the direction of maximum extension (X) within the plane of slaty cleavage itself.

extension lies within the plane of the slaty cleavage. This is known collo-quially as the stretching direction. It is often marked by the development of a fibrous mineral lineation, which can be seen by careful inspection of the surface formed by a slaty cleavage. Deformed objects such as rock fragments, detrital grains, concretions and oolites are all found with their long axes parallel to this lineation, which is therefore parallel to the direc-tion of maximum extension in the rock. Where the slaty cleavage strikes parallel to the trend of the orogenic belt, it is commonly found that the stretching direction plunges down its dip. This means that the stretching direction usually has a rather constant trend at right angles to the orogenic belt as a whole. It marks the upward stretching of the rocks as they are com-pressed between the margins of the orogenic belt, at right angles to its trend.

Cleavage Fans and Refraction.

A slaty cleavage is typically developed so that it is parallel to the axial planes of any folds, which affect only the bedding of sedimentary rocks in slate belts. However, such an axial-planar cleavage often forms convergent or divergent fans around the axial-planar surfaces, as shown in Figure 8.6A. This means that the cleavage is only parallel to this surface at the axial plane itself. What sort of cleavage fan is formed depends on the nature of the folding. Divergent fans are generally associated with Class 3 folds which are developed wherever incompetent rocks such as slates are inter-bedded with more competent layers. The slaty cleavage diverges from the axial plane as it is traced towards the core of a particular fold. Convergent fans are found to be associated with the Class 1 folds which are typically developed in the more competent layers. The slaty cleavage converges on the axial plane as it is traced towards the core of a particular fold.

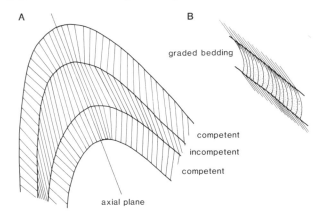

FIG. 8.6. A. Cleavage fans developed in a series of competent and incompetent beds, showing how the cleavage is refracted on passing from one bed to another. B. Effect of graded bedding on cleavage refraction.

Such changes in attitude of a slaty cleavage can result in its refraction as it is traced from competent to incompetent beds on the fold limbs. This occurs abruptly wherever there is a sharp contact between the two beds, while the slaty cleavage takes on a curved form if there is more gradual transition. For example, the slaty cleavage commonly shows an abrupt change in attitude as it cuts across the base of a graded bed, while it curves through the bed in such a way that it regains its original attitude towards the top, as shown in Figure 8.6B.

Relationship to Fracture Cleavage.

A slaty cleavage commonly changes in character as it cuts across the different beds in a fold. Thus, it may form a slaty fabric defined by the parallel orientation of layer-silicate minerals in slates, whereas it becomes dominated by the development of widely spaced mica films in the form of a spaced cleavage as it is traced into the more competent beds of greywackes or arkose. If these beds are orthoquartzite, the virtual lack of any layer-silicate minerals in the rock means that the mica-films are simply represented by a series of closely spaced fractures, arranged in the form of a convergent fan. This gives a structure known as fracture cleavage, which may be regarded as a form of slaty cleavage. Once formed, a fracture cleavage can be affected by shearing movements, so forming a series of faults on a small scale.

Bedding-cleavage Relationships.

Since a slaty cleavage is generally developed parallel to the axial planes of any folds in a slate belt, its presence can be used to determine the nature of these folds. It has already been noted that these folds affect only the bedding of the sedimentary rocks. Accordingly, it is the relationship of the slaty cleavage to the bedding which allows their nature to be determined.

It may be remembered that the axial plane is defined as the surface which passes through successive fold hinges in a series of folded surfaces. Moreover, the axial plane intersects each of the folded surfaces in a line which is parallel to the corresponding fold hinge. This means in effect that a slaty cleavage always intersects the bedding in a direction which is parallel to the fold hinge at a particular exposure, as shown in Figure 8.7. The bedding-cleavage intersection can be seen as a striping on the cleavage planes, formed by the slaty cleavage cutting across the lithological layers in the rock. This lineation is defined by the trace made by the bedding on the cleavage plane. Alternatively, if a bedding plane is exposed, the bedding-cleavage intersection is given by the trace made by the cleavage on the surface formed by the bedding plane. A fine ribbing is commonly developed on the bedding plane parallel to the bedding-cleavage intersection.

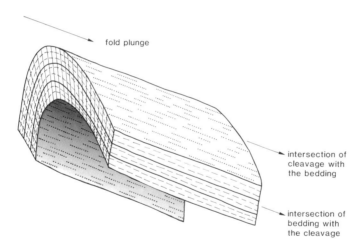

fold plunge

intersection of cleavage with the bedding

intersection of bedding with the cleavage

Fig. 8.7. Diagram showing how the bedding and cleavage intersect one another in a direction parallel to the plunge of the fold hinge in a cylindroidal fold.

Once the direction of the fold hinge has been determined from the attitude of the bedding-cleavage intersection, the relationship of the slaty cleavage to the bedding can be considered at a particular exposure. In general, it will be found that the slaty cleavage dips more steeply or less steeply than the bedding, while it may or may not dip in the same direction. Unless the slaty cleavage dips less steeply in the same direction as the bedding, the dip of the bedding can be used directly to determine the nature of the folding, as shown in Figure 8.8A. Thus, the bedding will dip towards the axial trace of a synform, while the axial trace of the complementary

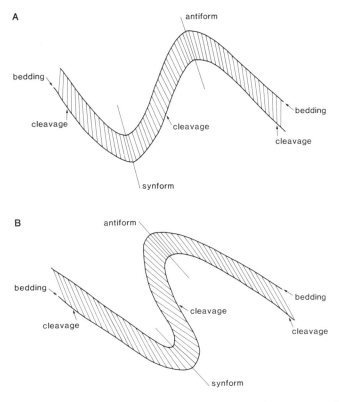

FIG. 8.8. Bedding-cleavage relationships in (A) upright folds and (B) overturned folds.

antiform would be encountered in the opposite direction. However, if the slaty cleavage dips less steeply in the same direction as the bedding, the exposure must be located on the overturned limb of a fold, as shown in Figure 8.8B. The slaty cleavage then dips towards the axial trace of an overlying antiform, while the axial traces of the underlying synform would be encountered in the opposite direction.

It should be noted that the dip of the slaty cleavage in either case would be parallel to the axial planes of the folds. Moreover, if the folds are not horizontal structures, the relationship of the slaty cleavage to the bedding should be studied by looking down the plunge of the bedding-cleavage intersection. This would show the structural relationships in the plane of the fold profile, at right angles to the fold hinges. It is important to realise that these relationships do not provide any information concerning the way-up of the rocks affected by the folding, unless the direction of structural facing is known. This means that the folds identified by the relationship of bedding to the slaty cleavage can only be described as antiforms and synforms.

Although these relationships are commonly developed in slate belts, it is now known that a slaty cleavage can form at an oblique angle to the axial planes of the associated folds. This is shown in the field wherever the intersection made with bedding by a particular cleavage plane crosses over an individual fold hinge. Such a relationship is known as transection. It gives rise to a non-cylindroided element in the structural pattern, since the bedding-cleavage intersections differ in orientation according to their structural position.

Structural Facing in Slate Belts.

If sedimentary structure are present which allow the way-up of the beds to be determined, the direction of structural facing can be found. Thus, the folds face upwards or downwards in the slaty cleavage, parallel to the direction in which younger beds are encountered at right angles to the bedding-cleavage intersection. This direction can be detemined wherever an exposure shows a slaty cleavage cutting across sedimentary structures such as graded bedding or cross bedding. Once the direction of structural facing has been determined, antiforms and synforms can be identified as anticlines or synclines, according to whether the structures face upwards or downwards.

Sedimentary structures can also be used to identify individual folds as anticlines or synclines, since they allow the stratigraphic order of the sedimentary rocks affected by the folding to be determined. It should be emphasised that anticlines are simply defined as folds with a core of older rocks, surrounded by younger rocks, while synclines are simply defined as folds with a core of younger rocks, surrounded by older rocks. Such a stratigraphic relationship does not identify a particular fold as an antiform or a synform, unless the form of the fold can be determined by other means. Even so, it is usually found that the structures developed within a slate belt are upward-facing, unless there is evidence for the development of recumbent folds and nappes.

Major and Minor Folds.

The discussion so far has assumed that folds are only developed on a particular scale within a slate belt, so that they all have the same wavelength. However, folding commonly occurs on different scales, depending on the thickness and spacing of the competent layers. A single layer embedded in less competent material will buckle with a wavelength that is directly proportional to its thickness. Likewise, a series of competent layers will act as a single multilayer, which buckles with a much greater wavelength than would otherwise be expected, provided that the individual layers are only separated from one another by a small amount of incompetent material. This means that any formation which consists mostly of competent rocks will buckle to form major folds on a large sclae, while the competent beds in any formation which consists mostly of incompetent rocks will buckle to form minor folds on a small scale. It is the minor folds which are observed in the field, while it is the major folds which control the outcrop pattern of the various formations.

Plunge of Minor Folds.

If the folding is cylindroidal, the hinges of the major and minor folds would be parallel to one another, as shown in Figure 8.9. Accordingly, by measuring the attitude of the minor fold hinges wherever they are seen in the field, the plunge and trend of the major fold hinges can be determined. This information could also be obtained by measuring the attitude of bedding-cleavage intersections wherever they are seen in the field, since these intersections are also parallel to the major fold hinges. Such relationships are simply a consequence of cylindroidal folding. This places a constraint on the orientation of the minor fold hinges and bedding-cleavage intersections, parallel to the major fold hinges. Such structures are said to be coaxial with one another. Otherwise, the folding would be non-cylindroidal in character.

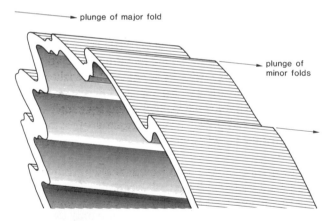

FIG. 8.9. Diagram showing how the plunge of minor folds gives the plunge of a major fold, provided that the folding is cylindroidal.

Whether or not the folding is cylindroidal can be determined from the orientation of the minor fold-hinges and bedding-cleavage intersections, as measured in the field. If these structures are not even locally coaxial with another, or if they show a gradual change in attitude as they are traced across a wider area, it can be concluded that the folding is non-cylindroidal on a major scale. However, it would be possible to divide the area into smaller sub-areas in the latter case, within which the folding could be regarded as cylindroidal.

Vergence of Minor Folds.

Another constraint is placed on the form of minor folds by the development of major folds. Thus, even if the major folds have fold limbs of equal length, the minor folds must have fold limbs of unequal length away from the major fold hinges. Figure 8.10 shows that two surfaces can be drawn tangential to a bed which is folded on a minor scale. They are known as enveloping surfaces, defining the sheet dip of the folded bed. These surfaces are parallel to the bedding where it is only folded on a major scale, so that they define the form of the major folds.

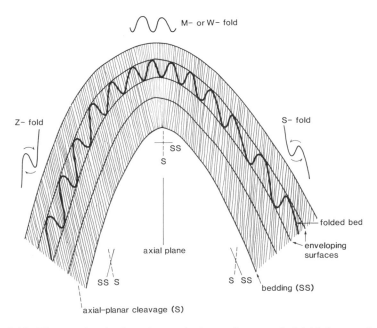

FIG. 8.10. Diagram showing how the enveloping surfaces to a bed folded on a minor scale define the form of the major folds affecting that bed. Note how the fold vergence and bedding-cleavage relationships change on crossing the axial plane of the major fold.

This means that the minor folds have a sense of asymmetry which depends on their position in relation to the major folds. For example, the minor folds developed near the hinge of a major fold tend to have fold limbs of the same length, since the enveloping surfaces are at right angles to their axial planes. Such folds are known as equant. However, as these minor folds are traced away from the hinge zone of a major fold, there is a gradual change in the relative lengths of their fold limbs. One set of fold limbs becomes relatively longer or shorter as the minor folds are traced in one or other direction, in comparison with the other set of fold limbs. Such folds are known as inequant. They are developed wherever the enveloping surfaces are inclined at an oblique angle to the axial planes of the folds.

Such a relationship implies that the sense of asymmetry shown by minor folds changes across the axial plane of a major fold. The asymmetry can be described in terms of the vergence shown by the minor folds. This term was originally introduced into the German literature to describe the fact that folds are inclined in a direction which is opposite to the dip of their axial planes. For example, a fold with an axial plane dipping to the south-east may be said to show north-westerly vergence. However, a separate term is not needed to describe this aspect of asymmetry. It can be specified in terms of the inclination, if this is simply defined as the direction opposite to the dip of the axial planes. Thus, a fold with an axial plane dipping towards the south-east would be inclined towards the north-west. This definition of inclination measures the asymmetry shown by a fold in relation to a horizontal plane. It is compatible with the description of folds as being overturned in a particular direction. Thus, folds overturned to the north-west would also be inclined in the same direction, opposite to the dip of their axial planes.

Vergence can then be used to describe the asymmetry of minor folds in relation to the dip of their enveloping surfaces. Thus, as viewed in the plane of the fold profile, minor folds have short limbs which appear to have rotated clockwise or anticlockwise in relation to the present attitude of their long limbs. The vergence of these folds is defined as the horizontal direction in the plane of fold profile, towards which the upper component of this apparent rotation is directed. So defined, the vergence shown is a pair of minor folds is directed towards the axial trace of a major antiform, while the axial trace of the complementary synform would be encountered in the opposite direction, as shown in Figure 8.11. The minor fold-pair can be said to verge towards the antiform, away from the synform.

This means that the axial traces of the major folds can be located by studying the changes in vergence shown by the minor folds, wherever they are observed in the field. The relationship of the bedding to the slaty cleavage can also be used, wherever the bedding is not affected by minor folding. It should be noted that the minor folds developed at the major fold-hinges have neutral vergence, since they form equant folds with limbs of equal length, while the bedding in a similar position lies at right angles to the slaty cleavage, if it is not affected by minor folding. These relationships allow the axial traces of the major folds to be located with more precision than would otherwise be possible from a casual inspection of the geological map.

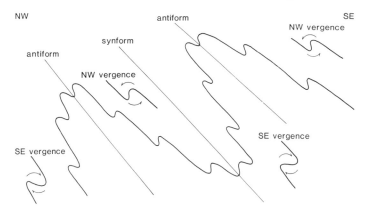

FIG. 8.11. Diagram showing how the vergence of minor folds can be used to map folds on a larger scale.

Effect of Deformation on Cylindroidal Folds.

The folds associated with a slaty cleavage are generally formed with horizontal hinges that trend parallel to the length of the orogenic belt. However, it is commonly found that the fold hinges are not all strictly parallel to one another, so that they are dispersed to some extent about the direction of a mean fold axis. Individual folds are likely to have curved hinges, which cannot be traced very far in the direction of the mean fold axis. The irregular nature of the folding tends to be accentuated by the deformation, so that cylindroidal folds are converted into non-cylindroidal folds, as shown in Figure 8.12A. How this happens can be considered in terms of the deformation associated with the development of a slaty cleavage.

It has already been mentioned that a slaty cleavage develops at right angles to the direction of maximum compression, as determined from deformed objects. Moreover, the stretching direction lying within the slaty cleavage marks the corresponding direction of maximum extension. Since the stretching direction plunges down the dip of the slaty cleavage, the strike of the slaty cleavage would be a direction which is generally affected by little or no deformation. This direction corresponds to the trend of the mean fold axis, at least approximately. Accordingly, it can be assumed that the plane of the slaty cleavage is simply affected by extension acting parallel to the stretching direction.

It can now be considered how the orientation of a fold hinge changes as a result of the stretching which occurs within the plane of the slaty cleavage. If such a fold hinge is perpendicular to the stretching direction, it will not be affected by any deformation. However, if it lies at an oblique angle to this direction, it will tend to rotate towards the stretching direction, away from

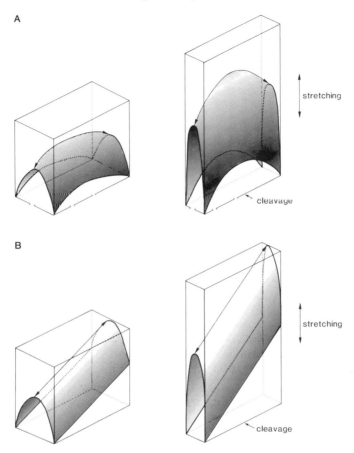

FIG. 8.12. Development of oblique folds as the result of deformation within the slaty cleavage.

its original position, while it undergoes a gradual increase in length, as shown in Figure 8.12B. Such structures are best termed oblique folds, wherever their fold hinges are not perpendicualr to the stretching direction.

It has already been mentioned that the individual hinges of cylindroidal folds would be dispersed about the direction of the mean fold axis. If the mean fold axis is approximately at right angles to the stretching direction, as shown in Figure 8.12A, the individual fold hinges will lie on either side of the stretching direction. They would then rotate in opposite directions, depending on the angle that they originally made with the stretching direction. This produces a series of oblique folds, pitching on either side of the stretching direction in the plane of the slaty cleavage. An individual fold with a curved hinge would be converted into a doubly-plunging antiform or synform. However, if the mean fold axis is not even approximately at right angles to the stretching direction all the individual fold hinges tend to rotate in the same direction, so reducing the angle that they originally made with the stretching direction, as shown in Figure 8.12B. This means that the original distribution of the individual fold hinges about the mean fold axis becomes increasingly asymmetrical to the stretching direction as the amount of deformation increases. There is a complete gradation between these two extremes, according to how the individual fold hinges were initially distributed in relation to the stretching direction. The development of oblique folds in this way may also be accentuated by variations in the intensity of the deformation along the trend of the orogenic belt, which are known to occur. Such variations in the intensity of the deformation would result in the development of folds with curved hinges, even if they were originally formed with straight hinges.

Orientation of Oblique Folds.

The stretching direction commonly has a very uniform attitude at right angles to the trend of an orogenic belt since it pitches steeply down the dip of the slaty cleavage. Although folds may be formed with their hinges parallel to the regional trend, continued deformation causes these fold hinges to rotate towards the stretching direction until they reach a position almost at right angles to their original attitude. This means that horizontal folds would be converted into vertical or reclined folds, depending on the dip of their axial planes, while individual folds often show strongly curved hinges passing through the horizontal at right angles to the stretching direction. However, it should be realised that the major folds may still retain an attitude which is parallel to the regional trend, at least overall, even although the minor folds appear to be oblique structures. This would be the case wherever a major fold was formed with its hinge originally at right angles to the stretching direction.

The oblique folds developed in response to stretching within their axial planes should be clearly distinguished from cross-folds, which are best defined to include all those folds formed with their hinges strictly parallel to the stretching direction. Such a distinction becomes more difficult to make as the intensity of the deformation increases, since the oblique fold hinges rotate towards the stretching direction until they are virtually parallel to this direction. However, it is often possible to determine than an individual fold hinge is slightly oblique to the stretching direction by careful inspection of the surfaces formed by the slaty cleavage in the field. It should be noted that the facing and vergence of these oblique folds remains constant in the original plane of the fold profile, parallel to the stretching direction. Such components of vergence and facing can therefore be used to determine the nature of the major folds.

Belts of Polyphase Deformation

The cores of orogenic belts are usually formed by rocks which have been affected by a polyphase history of deformation and metamorphism, even although such effects can also be seen in other tectonic settings. There is often sufficient evidence to suggest that the earliest phase in this history was marked by the formation of a slaty cleavage under conditions of low-grade metamorphism, which therefore extended throughout most of the orogenic belt prior to the onset of regional metamorphism at a higher grade. Thus, a slaty cleavage can often be traced into areas of more structural complexity, where it is seen to be affected by the development of later structures. Furthermore, even although its character tends to be obscured by the development of these structures under conditions of increasing metamorphic grade, it is commonly found that relicts of such a early, fine-grained fabric are preserved as inclusions within metamorphic minerals like garnet and feldspar. This means that the marginal zones of an orogenic belt are formed by rocks which have not been affected by subsequent events in the deformation history, while the central core of regionally metamorphosed rocks has been so affected. Since a slaty cleavage is typicaly developed axial planar to folds which affect only the bedding of the rocks its formation marks the first event in such a polyphase history of deformation and metamorphism.

Development of Crenulation Cleavages.

The margins of such belts of multiple deformation and metamorphism are generally marked by the gradual development of strain-slip or crenulation cleavages as secondary structures, superimposed on a slaty cleavage. The characteristic feature of such a crenulation cleavage is that the slaty cleavage is folded or crenulated on a very small scale to form a series of microfolds, as shown in Figure 8.4B. It is the axial planes of these microfolds which define the planes of crenulation cleavage in the rock. Such a cleavage is thus formed by discrete planes, along which the rock has a tendency to split. These planes are rarely more than a few millimetres apart, unless the rock is very coarse-grained. However, they may be so closely spaced that a careful inspection of a fractured surface is needed before their discrete nature can be observed.

The microfolds forming a crenulation cleavage can vary considerably in style from kink-like folds with angular hinges to rather curvaceous folds with rounded hinges. The form of these microfolds also varies according to how the crenulation cleavage planes are orientated in relation to the overall attitude of the layering which is affected by the microfolding. Equant microfolds are produced with limbs of equal length wherever the cleavage planes are roughly at right angles to the layering. However, inequant microfolds are produced with limbs of unequal lengths wherever the cleavage planes occur at an oblique angle to the layering, as commonly found to be the case.

The long limbs of such inequant microfolds may then be affected preferentially by the deformation. This is marked by the apparent development of shear zones, between which the slaty cleavage is sigmoidally folded. It is usually possible to trace the slaty cleavage at a low angle across these shear zones in such a way that they can be seen to develop directly from the long limbs of inequant microfolds. This process if often associated with a certain amount of metamorphic differentiation, since these shear zones become enriched in layer silicates and ore minerals while the intervening layers between these shear zones become enriched in quartz and felspar. These layers are generally known as microlithons. The shear zones separating these microlithons from one another are often affected by slight displacement, so that they may have the form of closely spaced microfaults.

Crenulation cleavages usually occur in single sets, superimposed on a slaty cleavage. The cleavage planes belonging to each set are all roughly parallel to one another. They typically occur parallel to the axial planes of folds which affect the slaty cleavage as well as the bedding. Such a crenulation cleavage shows a change in the sense of its displacement on passing from one fold-limb to another, as shown in Figure 8.13. Such folds become increasingly well developed as they are traced into the central parts of an orogenic belt, where the rocks have been affected by a polyphase history of deformation and metamorphism. However, crenulation cleavages can also occur as conjugate sets, lying at a high angle to one another. The two sets of cleavage planes often intersect the slaty cleavage at approximately the same angle, so that they are symmetrical. Such cleavages may be developed parallel to the axial planes of conjugate folds, which are generally developed on only a small scale.

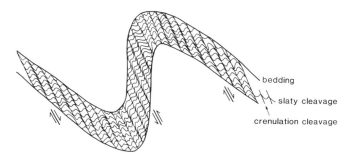

Fig. 8.13. Development of a crenulation cleavage as an axial-planar structure. Note change in displacement on the cleavage-planes in passing from one fold limb to another.

Evidence for Polyphase Deformation.

Crenulation and slaty cleavages resemble one another since they are both developed as axial-planar cleavages to folds. However, since a slaty cleavage affects only the bedding, its formation may reasonably be assumed to mark the first event in any polyphase history of deformaiton and metamorphism which affects the rocks under consideration. This means that a slaty cleavage or its equivalent in higher-grade rocks is axial planar to the earliest generation of folds which are developed in response to this history, unless there is any evidence to the contrary. Such a cleavage and folds are generally known as the first structures. For convenience, they are termed the S_1 cleavage and the F_1 folds, which are developed in response to the D_1 deformation, while the bedding is described as SS.

Since a crenulation cleavage affects a slaty cleavage as well as the bedding, its formation must be associated with a later event in the structural history of the area. This means that a crenulation cleavage is generally found axial planar to folds which belong to a later generation than the F_1 folds, unless there is any evidence to the contrary. If these structures developed in response to the second phase in the structural history, they are known as second structures. They can then be termed the S_2 cleavage and the F_2 folds, which were developed in response to the D_2 deformation. Likewise, the third structures would be represented by the S_3 cleavage and the F_3 folds, which were developed in response to the D_3 deformation, and so on.

Since different generations of structures occur in rocks affected repeatedly by deformation and metamorphism these structures can be dated with respect to one another, as shown in Figure 8.14. Thus, early cleavages are folded around the hinges of the later folds, while later cleavages cut across the axial planes of the earlier folds. The former relationship is seen wherever a slaty cleavage is folded around the hinge of a second or later fold, while the latter relationship is developed wherever a crenulation cleavage is superimposed on a slaty cleavage that is axial planar to a first fold. Such relationships are typically found wherever early folds have been affected by later folds on the same scale. The axial planes of the early folds are then folded, along with the bedding, to form the hinges of the later folds. This means that the bedding becomes refolded wherever there is more than one generation of fold structures. Refolding is therefore a characteristic feature of the central cores of orogenic belts, where the rocks have been affected by a polyphase history of deformation and metamorphism.

Slaty and crenulation cleavages can usually be distinguished from one another in areas of low-grade metamorphism, at least under the microscope, even if the fabrics are extremely fine-grained. However, such cleavages grade into what is known as a schistosity, which develops as the rocks become coarser-grained as a result of the recrystallisation and grain

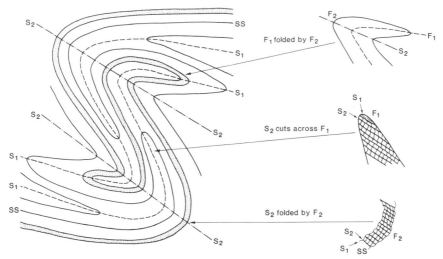

Fig. 8.14. Structural relationships used as evidence for dating folds and cleavages as belonging to different phases in the deformation history.

growth that occurs during any increase in the metamorphic grade. Such a schistosity often lacks any diagnostic features which would allow its relative age to be determined. In particular, the characteristic features shown by a crenulation cleavage tend to disappear as the micas in the rock recrystallise with a parallel orientation along both the discrete mica films and within the intervening microlithons. It may only be possible to date such a schistosity as relatively late if garnet or felspar porphyroblasts are present within the rock, containing discordant inclusion trails of an earlier fabric. Otherwise, there are no obvious features that would allow it to be distinguished from a schistosity formed from a slaty cleavage, at least in the field. It should be noted that such schistosities are characteristic of the earlier phases in the deformation history, since coarse-grained crenulation cleavages tend to be formed after the climax of progressive metamorphism in any particular area.

A characteristic feature of polyphase deformation is that the earlier generations of folds tend to be tight or isoclinal structures with a slaty cleavage or a coarser-grained schistosity parallel to their axial planes, while the later generations of folds are more open structures with a crenulation cleavage parallel to their axial planes. There is therefore a gradual decrease in the intensity of the deformation during the course of the deformation history. However, it is also the case that folds belonging to a particular generation become tighter and more compressed as they are traced towards the areas of high-grade metamorphism, forming the central core of an orogenic belt. This means that the earlier folds become increasing more difficult to recognise in these areas, where they often form isoclinal structures with a highly attenuated style. The bedding of the folded rocks would then appear to coincide with the schistosities that are developed parallel to the axial planes of these folds, giving rise to what has been termed a bedding-plane schistosity, often mistakenly.

The structural relationships developed in response to a polyphase history of folding are highly complex. However, the folds formed in response to a particular phase in the deformation history always share the same axial-planar cleavage as one another, even although the folding may occur on different scales. This places an important constraint on how structures of the same generation are related to one another. The structural relationships developed as a result have already been described with respect to the first phase in the deformation history. Similar relationships are found between the structures developed in response to subsequent phases in the deformation history, wherever a crenulation cleavage or schistosity is formed axial planar to folds belonging to the same generation but differing in scale from one another. However, the structural relationships so developed tend to be more complex since these structures affect rocks which have already been folded.

Orientation of the Second Folds.

It is commonly found that the folds formed during a particular phase in the deformation history have an axial-planar cleavage which is rather uniform in its attitude. The hinges of these folds are then simply defined by the direction in which this cleavage intersects the folded surface. For example, it has already been described how the F_1 fold hinges are parallel to the direction in which the S_1 cleavage intersects the bedding. However, the F_2 folding can affect the S_1 cleavage as well as the bedding. This means that the F_2 folds can be defined in two different ways.

If the F_2 folds are considered to affect the S_1 cleavage rather than the bedding, the F_2 hinges would be parallel to the direction in which the S_2 cleavage intersects the S_1 cleavage. Since the S_1 cleavage is likely to form with a rather uniform attitude, the F_2 hinges defined in this way are likely to plunge consistently in the same direction.

However, this is not the case if the F_2 folds are considered to affect the bedding rather than the S_1 cleavage, since the bedding varies in attitude as a result of the F_1 folding, as shown in Figure 8.15. The F_2 folds are then defined by the way that the S_2 cleavage cuts across the F_1 folds. This means in effect that the F_2 hinges defined by the bedding vary in attitude as they are traced across the F_1 folds. However, all the F_2 fold hinges along a particular line would then lie in the same plane, parallel to the S_2 cleavage.

Such variations in the attitude of the F_2 folds would not occur if the F_1 folds were isoclinal, so that the bedding and the S_1 cleavage were effectively parallel to one another on the F_1 fold limbs. Neither would such variations

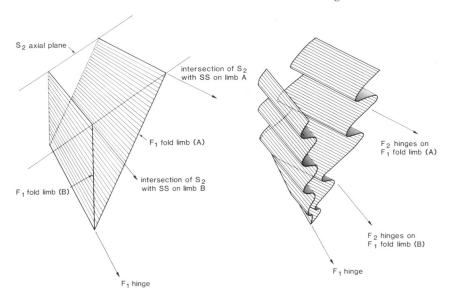

FIG. 8.15. Diagram showing how the form of an F_1 fold controls the attitude of F_2 fold hinges developed on its limbs (A) and (B).

be developed if the S_2 cleavage intersected the S_1 cleavage in a direction parallel to the F_1 hinges, since the F_1 and F_2 folds would then be coaxial with one another. Under these circumstances the S_2 cleavage also intersects the bedding in the same direction, so that the F_2 hinges are parallel to the F_1 hinges. This is the case whether the F_2 folds defined by the S_1 cleavage or the bedding. Although both are special cases, the first situation is perhaps the more likely, since the earlier phases in the deformation history commonly lead to the development of tight if not isoclinal folds whereas the later folding produces more open structures.

Orientation of the First Folds.

If the F_1 and F_2 folds were coaxial with one another, the F_2 folding would have no effect on the orientation of the F_1 hinges. The F_2 folding would then only affect the bedding and the S_1 cleavage which is developed parallel to the axial planes of the F_1 folds. However, since it is rather unusual for the F_1 and F_2 folds to be coaxial, the F_2 folding generally does affect the orientation of the F_1 hinges, as shown in Figure 8.16. How this happens is rather complex, since it depends on the nature of the F_2 folding. If the F_2 folds are parallel in style, the F_1 hinges tend to be folded around the hinges of the F_2 folds so that the angle between the F_1 and F_2 hinges remains effectively the same. This means that individual F_1 hinges tend to lie within a conical surface with its axis parallel to the F_2 hinges. By way of

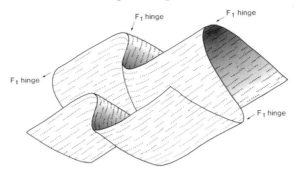

FIG. 8.16. Diagram showing the effect of F_2 folding on the orientation of F_1 fold-hinges.

contrast, if the F_2 folds are similar in style, the F_1 hinges tend to be folded around the hinges of the F_2 folds so that the angle between the F_1 and F_2 hinges changes. This occurs in such a way that the individual F_1 hinges always lie in the same plane, cutting across the axial planes of the F_2 folds.

Development of Non-cylindroidal Structures.

In general, non-cylindroidal folds are produced as a result of refolding unless the two generations of fold-structures are coaxial with one another, as shown in Figure 8.17. Thus, even if the earlier folds are cylindroidal in character, this feature is destroyed by later folding. Furthermore, the later folds are superimposed on the earlier structures, in such a way that they form non-cylindroidal structures. This means that non-cylindroidal folds are a characteristic feature of the central cores of orogenic belts, wherever the structural history embodies more than a single phase of folding.

The angular relationships so far described between F_1 and F_2 folds do not depend on any differences in scale that might exist between the two generations of fold structures. However, this factor does control the actual form of the geological structures which develop as a result of refolding. Three cases can be considered, depending on the relative difference in scale between the earlier (F_1) and later (F_2) folds. It should be realised that small-scale structures are generally developed as inequant structures with fold limbs differing in length. This means that the structural relationships are less complex than otherwise might be expected.

Firstly, the F_1 folds may be developed on a smaller scale than the F_2 folds. The earlier structures then exist as minor folds while the later structures occur as major folds. Under these circumstances, the earlier structures change in attitude as they are traced across the later structures. In particular, the minor F_1 hinges and the axial-planar S_1 cleavage are folded around the hinges of the major F_2 folds, as shown in Figure 8.18A. It is the

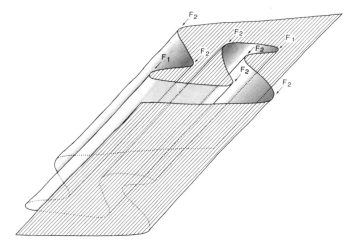

FIG. 8.17. Diagram showing coaxial F_1 and F_2 folds, giving rise to a cylindroidal structure.

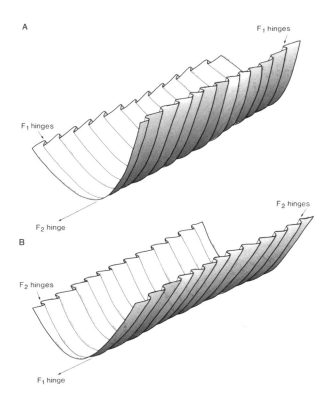

FIG. 8.18. A. Diagram showing minor F_1 folds affected by a major F_2 fold. B. Diagram showing minor F_2 folds superposed on a major F_1 fold.

changing attitude of the S_1 cleavage, or the bedding on the long limbs of the minor F_1 folds, which defines the form of the major F_2 folds. This means in effect that the major F_2 folds have a rather uniform plunge and inclination throughout the area under consideration. It is these folds which control the nature of the outcrop pattern, as defined by the geological boundaries between the various formations.

Secondly, it may be the F_2 folds which are developed on a smaller scale than the F_1 folds. The earlier structures then occur as major folds while it is the later structures which exist as minor folds. Under these circumstances, the minor F_2 folds are superimposed across the major F_1 folds with little effect on the overall structure. If it is assumed that the minor F_2 folds are defined by the bedding rather than the S_2 cleavage, they vary in plunge as they are traced across the major F_1 folds, as shown in Figure 8.18B. It is now the major F_1 folds which have a rather uniform plunge and inclination, so controlling the nature of the outcrop pattern. Even so, the bedding and the S_1 cleavage are affected locally by the minor F_2 folds.

Finally, the F_1 and F_2 folding may occur on approximately the same scale. This leads to the development of what are generally known as interference patterns, wherever the axial traces of the F_1 and F_2 folds cross one another. Three types of interference pattern are developed, according to how the F_1 and F_2 folds plunge in relation to one another. Such patterns can be developed on any scale. However, if they are developed on a sufficiently large scale, they can be recognised from the outcrop patterns shown by a geological map.

Development of Interference Patterns.

The first type of interference pattern is produced wherever the F_1 and F_2 folds are steeply inclined. The F_1 folds are then affected by the F_2 folds to form a series of domes and basins in the folded surfaces, as shown in Figure 8.19. Domes are formed wherever an F_1 antiform is crossed by an F_2 antiform while basins are formed wherever an F_1 synform is crossed by an F_2 synform. These domes and basins would be separated by saddles, wherever and F_1 antiform (or synform) is crossed by an F_2 synform (or antiform). The F_1 folds are so affected by the F_2 folding that they change in plunge across the axial traces of the F_2 folds. A culmination is formed wherever an F_1 fold hinge crosses an F_2 antiform, while a depression is formed wherever an F_1 fold hinge crosses an F_2 synform. Likewise, the F_2 folds are superimposed on the F_1 folds so they change in plunge across the axials traces of the F_1 folds. A culmination is formed wherever an F_2 fold hinge crosses an F_1 antiform, while a depression is formed wherever an F_2 fold hinge crosses an F_1 synform. This means that the F_1 and F_2 fold hinges plunge in opposite directions as the equivalent structures cross one another

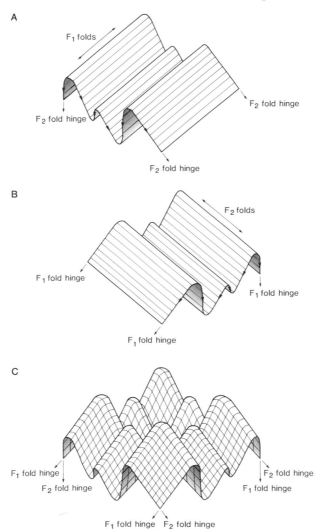

FIG. 8.19. Diagram showing development of domes and basins in a folded surface. A: Form of the F_1 folds. B: Form of the F_2 folds. C: Surface formed by the super-position of F_2 folds on F_1 folds.

to form a series of domes and basins in the folded surfaces. Such an interference pattern is shown in Fig 8.20A.

The second type of interference pattern is formed wherever the F_1 folds are overturned. The F_1 folds may still be affected by the F_2 folding so that they change in plunge across the axial traces of the F_2 folds. However, since the F_1 folds are overturned, the F_2 folds would be superimposed on these

A

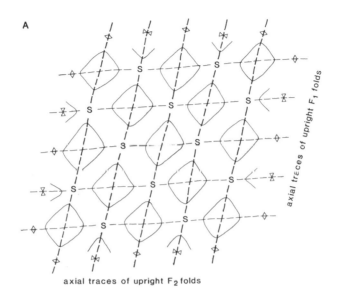

axial traces of upright F₂ folds

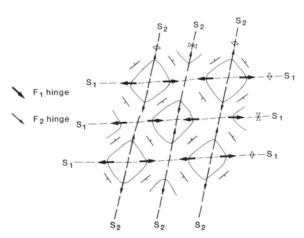

FIG. 8.20. A. Type 1 interference pattern in the form of domes and basins. S marks saddles in the folded surfaces. B. Type 2 interference pattern in the form of re-entrant domes and basins. Plunge arrows with hooks represent folds that have passed through the vertical to reach their present attitude, corresponding to a position on an overturned F₁ fold limb.

FIG. 8.20. Continued.

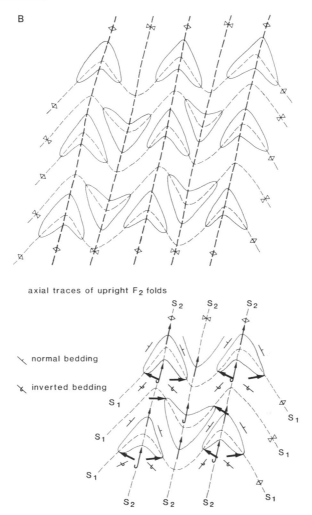

axial traces of upright F₂ folds

structures in such a way that the F_2 fold hinges pass through the vertical to plunge in the same direction as they are traced across the F_1 folds. Even so, culminations and depressions would still be developed along the F_2 fold hinges. This means in effect that a dome would be formed wherever a F_1 antiform is crossed by an F_2 antiform while a basin is formed wherever an F_1 synform is crossed by an F_2 synform. However, since the F_1 fold hinges plunge in the opposite direction while the F_2 fold hinges plunge in the same direction, wherever the F_1 and F_2 folds cross one another in this way, the domes and basins are marked by re-entrants. Such an interference pattern is

shown in Figure 8.20B. Typically the individual domes and basins form outcrops which have the shape of boomerangs. These outcrops can coalesce
with one another along the axial traces of either the F_1 folds or the F_2 folds,
to form very complicated patterns, as shown in Figure 8.21.

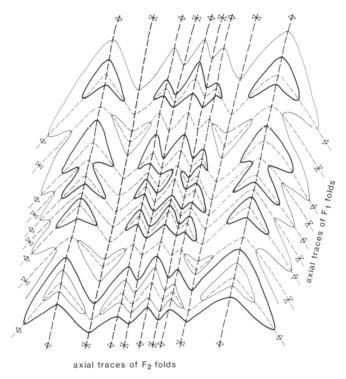

axial traces of F_2 folds

Fig. 8.21. Type 2 interference patterns formed by F_1 and F_2 folding on different scales.

The third type of interference pattern differs from the others since it is
not marked by the development of domes and basins wherever the F_1 and
F_2 folds cross one another. In particular, the F_1 folds do not change in
plunge as they are traced across the F_2 folds. This means that culminations
and depressions are lacking along the F_1 fold hinges. Instead, the F_1 folds
always plunge in the same direction, if the effect of the F_2 folding is taken
into account. The F_2 folds are then superimposed on the F_1 folds to give an
interference pattern which might be considered as typical of refolding. Such
an interference pattern is shown in Figure 8.22. It can be developed if the F_1
and F_2 folds are coaxial, even although this need not be the case. However,
if the F_1 and F_2 folds are not coaxial, the F_2 fold hinges form a series of
culminations and depressions as they are traced across the F_1 folds, while
the F_1 fold hinges vary in plunge as they are traced across the F_2 folds.

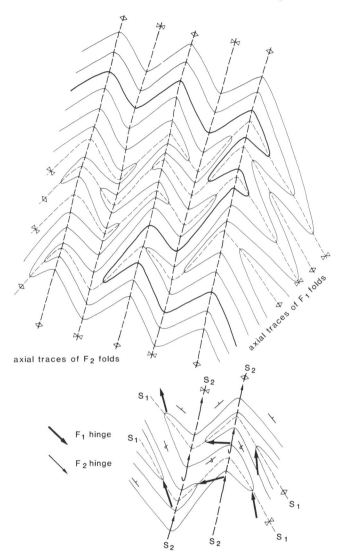

axial traces of F_2 folds

axial traces of F_1 folds

FIG. 8.22. A. Type 3 interference patterns developed by F_2 refolding of F_1 folds. B. Orientation of individual fold hinges.

Migmatite Complexes

The regional metamorphism of the rocks forming the central core of an orogenic belt is commonly accompanied by the development of what are known as migmatites, if the metamorphic grade becomes sufficiently high. Migmatites can best be defined as heterogenous rocks of composite origin, formed by the intermingling of granitic material with a metamorphic host.

Such rocks occur in the contact aureoles of granitic batholiths, particularly if they are intruded as concordant bodies during regional metamorphism, while they are also developed on a regional scale in the form of a migmatite complex. It is a characteristic feature of most migmatites that their formation is accompanied by the highly plastic folding and disruption of the metamorphic host rocks.

The structural features developed in response to migmatisation are highly complex, as can be seen from Figure 8.23. However, a number of different

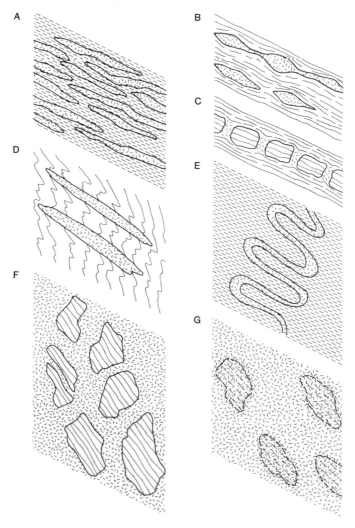

Fɪɢ. 8.23. Relationships of grantitc material (dashed ornament) to host rocks in migmatiles. See text for details.

rock types can be recognised. It is commonly found that veins and irregular sheets of granitic material are developed along the lithological layering of the rocks, so forming lit-par-lit gneisses (see Fig. 8.23A). The layering may be folded, along with these granitic sheets, by a subsequent phase in the deformation history. Irregular masses may also be developed as segregations of granitic material, which may be roughly aligned along the layering and which may be affected by pinch and swell (see Fig. 8.23B). Irregular segregations of granitic material can also be formed in areas of reduced pressure, wherever there is a tendency for cavities to open in the host rocks as the result of boudinage (see Fig. 8.23C). It is commonly found that the initial stages in the development of migmatites are marked by the widespread occurrence of granitic, pegmatitic and aplitic veins, cutting the metamorphic rocks at the margins of a migmatite complex. Such veins can also be developed within the complex itself, where they are often deformed along with their host rocks. These veins are commonly folded in a ptygmatic fashion (see Fig. 8.23E), while they may be developed along the axial planes of individual folds in the host rocks, as shown in Figure 8.23D. The latter circumstance suggests that they could then be dated as contemporaneous with these structures.

The granitic material in a migmatite commonly occurs as large feldspar porphyroblasts, or as quartzo-feldspathic segregations, around which the schistosity of the host rock is deflected, so forming what is known as an augen gneiss. Such rocks grade into permeation gneisses, wherever the feldspathic material has a more uniform distribution throughout the rock. Augen gneisses and permeation gneisses are usually formed from rocks which have such a composition that they can easily be affected by migmatisation. However, there are other rocks which tend to resist migmatisation. It is commonly found that these rocks respond by fracturing and disruption to the influx of granitic material, so forming a rock known as an agmatite. This has the appearance of a plutonic breccia, in which the host rocks are penetrated and disrupted by an irregular network of granitic veins, as shown in Figure 8.23F.

Although there may be relatively sharp boundaries between the granitic material and the metamorphic host in a migmatite, it is commonly found that these boundaries are rather indistinct. This tendency is enhanced wherever the host rocks have been affected by permeation to such an extent that they take on a granitic composition. This is marked by a gradual transition into a rock known as a nebulite, in which the amount of granitic material has so increased while the lithological identity of the metamorphic host has become so obscured by permeation that a clear distinction can no longer be made between these two components of the rock. This transition is marked by the presence of irregular and nebulous relicts of the modified host rocks, which often occur as wispy schlieren in otherwise homogeneous gneisses, as shown in Figure 8.23G.

Tectonic Slides

The metamorphic rocks forming the core of an orogenic belt are commonly cut by structural discontinuities in the form of tectonic slides. Such structures are faults, formed as a result of deformation under metamorphic conditions. It is commonly found that they are closely related to the major folds which formed during one or other phase in the deformation history. Thus, slides are often developed parallel to the axial planes of major folds in such a position that they occur on a particular fold limb, as shown in Figure 8.24.

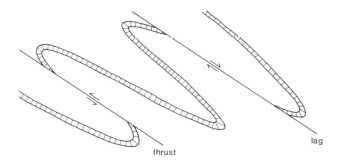

FIG. 8.24. Definition of tectonic slides as thrusts or lags.

They may then be defined as thrusts or lags, wherever the folds are inclined from the vertical. A slide would be recognised as a thrust if it was developed on the underlying limb of an early (F_1) anticline or a later (F_2) antiform, whereas it would be recognised as a lag if it was developed on the underlying limb of an early (F_1) syncline or a later (F_2) synform. Such definitions are based on the assumption that the adjacent folds increase in amplitude as a result of the sliding. Thus, thrusts are formed by a forward movement of the overlying rocks, whereas lags are associated with the reverse process. It is usually found that slides are formed during the earlier phases in the deformation history as a result of such movements.

Slides are not always related in this way to folds. Some slides are associated with disharmonic folds so that they act as planes of décollement, while other slides cut across the axial planes of major folds in such a way that they appear to be later than these structures. They may also be developed so that they are strictly parallel to the bedding of the rocks. However, since slides are generally formed relatively early in the deformation history, so that they are often associated with the F_1 folds, it is commonly found that they are folded along with bedding during the later phases in the deformation history.

It is characteristic feature of slides that they are often rather difficult to recognise in the field. This difficulty arises since most slides are developed in apparent conformity with the bedding on the limbs of isoclinal folds, while they are rarely marked by any brecciation or cataclasis of the adjacent rocks, unless this has occurred subsequently as a result of faulting. Slides often have the appeareance of a bedding plane under these circumstances. Admittedly, there is often a gradual increase in the intensity of the deformation towards the slide-plane. This would be marked by the attenuation of the bedding as a result of the deformation, while any cleavages or foliations associated with the sliding would become much more intensely developed in its vicinity. However, such evidence of extreme deformation is not diagnostic of a slide for two reasons. Firstly, some slides are not accompanied by an increase in the amount of deformation affecting the adjacent rocks, since it is possible for all the movements to be concentrated on the slide plane. Secondly, even if there is evidence for extreme deformation, this may not be accompanied by the development of a structural discontinuity within the zone so affected. This means in effect that slides can often only be recognised on stratigraphic evidence.

Slides and their Recognition.

Slides can best be recognised on the scale of a geological map wherever tectonic movements associated with the folding appear to have removed parts of the stratigraphic sequence. Typically, this is shown by different formations coming into contact with one another as they are traced throughout the region. A particular formation could then be recognised as occurring in contact with more than two other formations, so that the original juxtaposition of the stratigraphic formations is lost. Such cross-cutting relationships may be seen along the strike, or they may be reconstructed as the stratigraphic sequence is traced across the strike. They can only be developed if the apparent conformity shown by most slides in the field is an illusion.

Slides can only be recognised in this way if it is reasonable to assume that the stratigraphic sequence originally extended without much change across the whole area. This assumption may reasonably be made wherever the stratigraphic discontinuities in the outcrop pattern can be directly related to the presence of major folds, so that they always occur in a particular position in relation to the regional structure. This means that a clear understanding of the regional structure is required in order to establish that slides do occur in a particular area. However, it should be recognised that such discontinuities in the outcrop pattern may be the result of lateral changes in facies, which cause individual formations to disappear by thinning to zero, or that they may be related to the development of angular unconformities within the stratigraphic sequence, which can also cause individual formations to disappear as a result of overstep and overlap. If such discontinuities do have a stratigraphic origin, it is unlikely that they would be closely related to the development of the regional structures in the same way as tectonic slides.

Structural Features in Slides Zones.

Although there may be little structural evidence to distinguish a slide plane from a normal stratigraphic contact, many slides are marked by structural features that develop in response to extreme deformation of the adjacent rocks. Such a zone of highly deformed rocks is generally known as a slide zone. Since the rocks on either side of a slide zone are less deformed, they must have been displaced relative to one another as the result of shearing movements, in the manner of a fault. It is these movements which cause the bedding of the adjacent rocks to become attenuated with the slide zone, while they also cause the cleavage or foliation associated with the slide zone to become much more pronounced. This is accompanied by the bedding rotating into parallelism with the cleavage or foliation, or vice versa, so forming a platiness parallel to the margins of the slide zone. This platiness is analogous to a mylonitic banding, even although it is generally formed under more ductile conditions. The development of such a platiness is often associated with a pronounced stretching of the rocks within the slide zone, so forming a marked linear fabric parallel to the stretching direction.

The minor folds formed during the same phase in the deformation history would be expected to tighten as they are traced into the slide zone, where they may become so compressed and attenuated that they cannot easily be recognised. It is the cleavage axial-planar to these folds which becomes much more pronounced within the slide zone. The increase in the intensity of the deformation also causes the hinges of these folds to rotate towards the stretching direction, so forming oblique folds. If the deformation is sufficiently intense, these folds may appear to be parallel to the stretching direction. However, it is commonly found that folding affects the layering developed within the slide zone in such a way that folds are formed with their hinges strictly parallel to the stretching direction. These structures are termed cross folds, in distinction to oblique folds. Cross folds tend to be extremely cylindrical structures, since their fold hinges are parallel to the stretching direction, which has itself a very constant orientation. However, cross folds often have rather irregular profiles, since they lack well-defined axial planes in many cases.

Thrust and Fold Nappes

The margins of orogenic belts are often defined by overthrusts, which form a whole series of allochthonous thrust-sheets that have moved over the autochthonous rocks of the foreland. Similar relationships are developed within orogenic belts wherever sedimentary rocks that were originally deposited side by side are now found on top of one another. This can occur as a result of low-angle thrusting or recumbent folding, as shown in Figure 8.25. The structure formed as a result is known as a nappe, which is the French word for a sheet. It is a term that can be applied to any sheet-like mass of allochthonous rocks, which has been displaced horizontally in relation to the underlying rocks. The nature of this displacement may be rather uncertain. However, it must result in the nappe forming a tectonic cover to the underlying rocks. Nappes fall into the class of major structures, so that a considerable amount of displacement should have taken place for the term to be applied in this way.

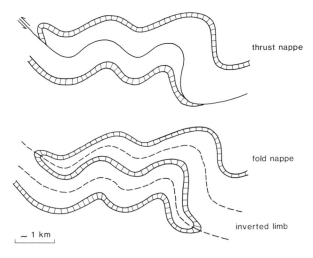

thrust nappe

fold nappe

inverted limb

~ 1 km

FIG. 8.25. Development of thrust and fold nappes.

Thrust nappes are developed wherever the sheet-like mass of allochthonous rocks has been displaced in relation to its substratum on a thrust plane. The rocks forming a thrust nappe may or may not be folded. Thrust nappes can be contrasted with fold nappes, which are formed as a result of recumbent folding on a regional scale. Slides are commonly developed in association with fold nappes, which they separate from one another. All these structures are generally formed during the early phases in the deformation history, so that they would be affected by the subsequent phases of folding.

Structural Features of Nappes.

It is often the case that thrust nappes rest on slides forming a tectonic contact in apparent conformity with the bedding, even although the rocks on either side may be tightly or even isoclinally folded. Likewise, fold nappes are usually associated with the development of isoclinal folds, so that any slides present are likely to appear parallel to the bedding of the rocks. Under these circumstances, the original sequence of sedimentary formations becomes so affected by the tectonic movements that an entirely new layering is produced. Thrust nappes are often marked by the tectonic repetition of the same sequence of sedimentary formations in apparent conformity with one another, while they can equally well be marked by the overthrusting of older sedimentary sequences on top of younger rocks to form an apparently continuous sequence of sedimentary formations. Similar features are found wherever slides are associated with the development of fold nappes, which are otherwise marked by the tectonic repetition of inverted sequences of stratigraphic formations, separated from one another by right-way-up rocks of the same age. The displacements associated with thrust nappes can be sufficient to bring stratigraphic sequences of the same age into juxtaposition with one another, even although they may show marked differences in sedimentary facies. Finally, it is possible for basement rocks to be involved in nappe structures, forming thrust sheets or anticlinal fold-cores as the case may be.

Reworking of Basement Rocks

So far, the structures developed within orogenic belts have only been considered as they affect the sedimentary cover rocks. However, these rocks are generally found to occur above an underlying complex of basement rocks, from which they were originally separated by a pronounced unconformity. This basement may retain a highly complex structure, which developed in response to an earlier history of deformation, metamorphism and igneous intrusion at considerable depths within the earth's crust. It is often termed the crystalline basement because it is composed of coarse-grained plutonic rocks known as gneisses.

The unconformity separating this basement complex from the overlying cover rocks marks an erosion surface, on which the overlying rocks of the cover sequence were deposited. It can therefore be conclude that it was originally horizontal or nearly so, forming a surface of low relief. There would be a marked discordance between the structures developed in the underlying basement and the bedding of the overlying rocks.

Effect on the Basement-cover Unconformity.

It is generally found that the unconformity forming the stratigraphic base of the cover rocks dips towards the margin of the orogenic belt as it is traced away from the craton. This means that the cover rocks become thicker in the same direction. Such an increase in thickness is likely to be an original feature, even although it may be emphasised by the tectonic effects of folding and overthrusting within the marginal zones of the orogenic belt itself. This is often accompanied by décollement of the cover rocks along bedding-plane thrusts, which may follow the basement-cover unconformity or a slightly higher horizon within the cover sequence. The basement rocks would then extend as a rigid mass for some distance below the deformed cover rocks, before they are affected by the orogenic movements.

The intital stages in the distortion of the basement-cover unconformity within an orogenic belt are usually marked by the development of thrust wedges, as shear zones start to form in the underlying rocks of the basement. These zones then act as the boundaries to wedge-like masses of basement rocks, which penetrate upwards into the overlying cover rocks. The basement rocks forming these upthrust masses are often separated from one another by narrow synclines, formed by the overlying cover rocks where these are highly compressed between the basement massifs.

The form of these structures becomes more extreme as they are traced towards the interior of the orogenic belt, where the deformation and metamorphism is sufficiently intense for the basement rocks to behave mechanically like the cover rocks. The basement-cover unconformity can be so affected by the deformation that it outlines the form of extensive thrust sheets and isoclinal fold-cores, associated with the development of nappes in the basement and over-lying cover rocks. The basement rocks forming these structures may extend for tens of kilometres from their roots within the orogenic belt. This illustrates the extreme nature of the deformation that occurs within orogenic belts.

Development of Mantled Gneiss Domes.

The basement-cover unconformity can also be deformed in a very different way to form what are known as mantled gneiss domes. These structures are produced as a result of vertical movements, which are most likely to be generated by gravitational instability. This would occur if the basement was formed by granitic gneisses that were less dense than the overlying cover rocks. There would then be a tendency for the basement rocks to rise diapirically into the cover rocks, forming dome-like masses on a fairly large scale. Individual domes may occur in a particular position because they develop from granite plutons in the underlying basement, even although this need not always be the case. It is commonly found that mantled gneiss domes form linear belts, parallel to the trend of the orogenic belt as a whole.

Mantled gneiss domes are similar in form and structure to concordant intrusions of granitic rocks, which were emplaced as diapirs. However, the cover rocks are younger not older than the granitic rocks forming the dome itself, from which they are separated by the basement-cover unconformity, as shown in Figure 8.26. The cover rocks may contain detritus which has been derived from the erosion of the underlying rocks of the granitic basement. They may also be intruded by components of the basement complex, forming dykes and apophyses of granitic rocks. Although these intrusions must be younger than the cover rocks, they appear to be derived from the underlying basement rocks through the action of regional metamorphism and migmatisation.

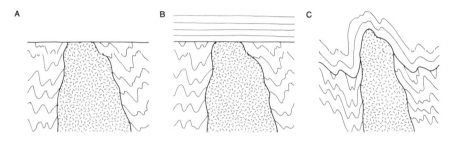

FIG. 8.26. Stages in the formation of a mantled gneiss dome (after Eskola). (A) Intrusion of granite pluton into folded country rocks, followed by uplift and erosion. (B) Deposition of overlying sedimentary sequence. (C) Mobilisation of granitic pluton to form mantled geiss dome.

Reactivation and Mobilisation.

New structures are formed in the basement rocks as they become affected by the deformation and regional metamorphism associated with the development of the orogenic belt, as shown in Figure 8.27. Initially, this is marked by the formation of discrete shear zones that cut across the pre-existing structures of the basement rocks, parallel to the margins of the orogenic belt. A mylonitic banding commonly develops within these shear zones by the extreme attenuation of the various components within the basement complex. Traced towards the orogenic belt, these shear zones widen as they start to coalesce with one another. This is accompanied by the mylonitic banding grading into a foliation, which tends to obliterate the pre-existing structures of the basement rocks. Finally, regional metamorphism and migmatisation has the effect of reconstructing these rocks in an entirely new form. Such processes of tectonic and metamorphic regeneration, which affect the basement rocks deep within an orogenic belt, are know as reworking or reactivation. They may be accompanied by mobilisation, wherever the rocks are intruded by granitic and pegmatitic material which has been derived by partial melting or hydrothermal activity from within the basement itself. However, it should be realised that such granitic material now found within these reactivated basement complexes may be derived from the mantle. Reworking has the effect of forming a structural

FIG. 8.27. Effect of reworking on the structural features of basement rocks at the margin of a orogenic belt.

and metamorphic front against the basement rocks which form the foreland to the orogenic belt, beyond which they retain their original features.

The reworking of a basement complex wihin an orogenic belt has the effect of gradually obliterating the angular discordance which originally existed between the structure of the basement rocks and the bedding of the cover rocks. It is found that this discordance eventually becomes impossible to recognise as it is traced into the orogenic belt, where the basement rocks become intensely deformed along with their sedimentary cover. As already mentioned, this may be associated with the development of thrust and fold nappes, so that the basement rocks form thrust sheets and isoclinal fold-cores which are tectonically inserted into the cover sequence. Basement and cover then appear to share the same structure, since the pre-existing structures of the basement rocks have rotated into parallelism with the bedding of the cover rocks to form a tectonic layering. It is often very difficult to determine the nature of the early structures which have produced this layering. This difficulty arises partly not only because the early structures tend to be extremely compressed and attenuated but also because the layering is affected by the subsequent folding and refolding during the later phases in the deformation history.

Effect on Dyke Swarms.

The clearest evidence for reworking is seen wherever the basement complex is cut by a swarm of basic bykes, which were intruded prior to the subsequent formation of an orogenic belt. It is these dykes which may be used to define the structure of the basement rocks. Reworking then has the effect of reducing the angular discordance between these dykes and the cover rocks until it is entirely lost. Likewise, if the dykes are discordant structures cutting across the foliation and layering in the basement rocks, it is found that reworking gradually reduces this discordance to zero, as shown in Figure 8.28. The effect of reworking on the individual dykes within the swarm depends on their relative orientation. They tend to fold wherever they are compressed within their own plane, while they tend to become disrupted into discrete fragments if they are stretched parallel to

Fig. 8.28. Effect of reworking on an discordant dyke cutting basement gneisses.

this direction. Such changes are generally accompanied by recrystallisation and the growth of new minerals under the changed conditions of temperature and pressure within the orogenic belt, while individual dykes may be intruded by veins of granite, pegmatite and aplite. It is therefore difficult to trace these dykes as mappable units into the orogenic belt for any distance, once their dyke-like structure is destroyed by reworking.

Relative Dating of Orogenic Belts

So far, only the structural features displayed by orogenic belts have been considered. However, since the continental crust is typically formed by a whole series of orogenic belts which developed during the course of geological history, these belts can be dated in relation to one another. How this is done depends to a considerable extent on whether it is the folded cover rocks or the reactivated basement rocks which are exposed by an orogenic belt at the present level of erosion.

If it is the folded cover rocks which are exposed in this way, it is usually possible to trace these rocks across the foreland of the orogenic belt, away from its margins. They would eventually be found to rest unconformably as a flat-lying sequence of sedimentary rocks on top of older rocks. The structures developed within these older rocks might well define the outlines of another orogenic belt, which must itself be the older of the two belts. This conclusion does not require the older belt to be truncated by the younger belt, even although this might well be the case.

Alternatively, it might be reactivated basement rocks which are exposed within the confines of an orogenic belt. This usually means that the rocks exposed beyond the limits of the orogenic belt are represented by basement rocks which have not been affected by reworking, so that they preserve their pre-existing structures. It is these structures which are truncated along the margins of the orogenic belt, which must therefore be younger in age. It is generally found that reworking produces new structures that trend parallel to the margin of the younger orogenic belt in such a way that they cut across the older structures of the basement complex, as shown in Figure 8.29. This is most clearly seen where these older structures are simply defined by a swarm of basic dykes. However, they can be defined by the structural trends developed within the basement rocks during an earlier tectonic cycle. It is these pre-existing structures which are truncated by the structural and metamorphic front forming the margin of the younger orogenic belt.

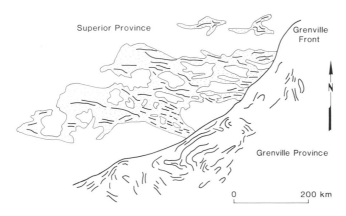

FIG. 8.29. Truncation of Archaean structures in the Superior Province by the younger Proterozoic structure of the Grenville Province in the Canadian Shield (after the 1:1,500,000 International Tectonic Map of North America).

These stratigraphic and structural relationships can be used to date the orogenic belts which have been formed in succession by different tectonic cycles during the course of geological history. Each tectonic cycle produces an orogenic belt where the cover rocks are folded and metamorphosed and where the underlying basement rocks are affected by reworking. The cover rocks typically extend outwards beyond the margins of the orogenic belt to form a flat-lying sequence of sedimentary rocks, which rest unconformably on a basement of older rocks. However, this basement consists of rocks which have been affected by orogenic movements during earlier tectonic cycles. It is therefore possible to recognise a whole series of orogenic belts within these basement rocks, which can also be dated in relation to one another. Each belt consists of folded and metamorphosed cover rocks, which originally rested unconformably on a basement of older rocks, even although this basement has now been affected by reworking. The cover rocks extend outwards from the orogenic belt to rest unconformably on a basement of even older rocks, retaining the structural features developed during earlier tectonic cycles.

Tectonic Provinces in the Continental Crust.

This means in effect that the continental crust can be divided into a series of tectonic provinces, marking the successive development of orogenic belts during the course of geological history. The oldest provinces that can clearly be recognised within the Precambrian shields were affected by a long and very complex history of geological events during the Archaean. The rocks within these Archaean provinces retain the structural and other features which developed in response to this history, since they have not been affected by subsequent reworking during later tectonic cycles. It can therefore be concluded that these parts of the continental crust became stable at the end of the Archaean, even although they had previously been affected by repeated episodes of deformation, metamorphism and igneous intrusion.

Since these areas were the first to become tectonically stable, they can be regarded as continental nuclei. However, this does not imply that they represent the entire extent of continental crust at the end of the Archaean. Since the structural trends developed within these Archaean provinces are truncated by the younger orogenic belts around their borders, these Archaean rocks must have originally extended over much wider areas than they do now. This means that Archaean rocks can often be recognised to form the reactivated basement rocks within tectonic provinces of a younger age. The limited extent of Archaean rocks at the present day cannot therefore be taken as evidence for continental growth by the accretion of successive orogenic belts.

These Archaean provinces are now seen to be surrounded by a whole plexus of orogenic belts, which became stabilised at different times during the Proterozoic and the Phanerozoic when particular cycles of tectonic activity came to an end. Obviously, it is the orogenic belts that became stabilised at the end of the Early Proterozoic which would have encircled the Archaean provinces. However, subsequent cycles of tectonic activity can produce orogenic belts cutting across the boundaries between the tectonic provinces that were formed by the stabilisation of earlier orogenic belts. Although such belts are quite common, other belts are found to surround the pre-existing provinces so that they become progressively younger in age as they are traced away from the continental nuclei.

As orogenic belts become stabilised at the end of particular cycles of tectonic activity, the cratonic areas formed as continental nuclei at the end of Archaean would become larger and larger. If an older tectonic province was flanked by a younger orogenic belt, the cratonic area formed by this province would simply increase in size once the orogenic belt became stabilised at the end of its development. Alternatively, if two tectonic provinces are separated from one another by a younger orogenic belt, the cratonic areas formed by these provinces would coalesce with one another to form a single cratonic area of larger size, once the orogenic belt became stabilised at the end of its development. The cratonic areas in existence at the end of a particular tectonic cycle would then act as the forelands to the orogenic belts which developed in response to the next cycle of tectonic activity.

This means that the geological history of the continental crust appears to be marked by the progressive enlargement and amalgamation of cratonic areas at the expense of orogenic belts. However, this tendency is likely to undergo local and temporary reversals during the course of geological history for two reasons. Firstly, orogenic belts may well become established so that they affect cratonic areas which have previously been stabilised at the end of an earlier tectonic cycle. This would lead to a relative increase in the area of continental crust affected by orogenic movements, at the expense of the intervening cratons. It has already been emphasised that this must occur wherever orogenic belts are seen to cut across one another, at least to a limited extend. Secondly, this may result in an orogenic belt cutting across the craton in such a way that it becomes divided into two parts, which would then be recognised as separate cratons. This would lead to an increase in the number of cratonic areas with the passing of geological time, even although each area would be much reduced in size. Such changes may be the result of plate tectonic movements in the remote past, leading to the fragmentation and reassembly of the continental crust to form a series of cratonic areas, separated from one another by orogenic belts at any particular time.

Selected References on Structural Geology and Tectonics

Allum, J. E. A., (1966) *Photogeology and Regional Mapping*. Pergamon, Oxford.

Anderson, E. M., (1951) *The Dynamics of Faulting* (2nd Edition). Oliver & Boyd, Edinburgh.

Badgley, P. C. (1959) *Structural Methods for the Exploration Geologist*. Harper, New York.

Badgley, P. C. (1965) *Structural and Tectonic Principles*. Harper, New York.

Bailey, E. B. (1935) *Tectonic Essays, Mainly Alpine*. Clarendon Press, Oxford.

Billings, M. P. (1972) *Structural Geology* (3rd Edition). Prentice Hall, Englewood Cliffs.

Bishop, M. S., (1960) *Subsurface Mapping*. Wiley, New York.

Blackadar, R. G. (1960). *Guide for the Preparation of Geological Maps and Reports*. Miscellaneous Report 16, Geological Survey of Canada, Ottawa.

Braunstein, J. and O'Brien, G. D. (Editors (1968) *Diapirs and Diapirsion*. Memoir 8, American Association of Petroleum Geologists, Tulsa.

Bush, H. G. (1929) *Earth Flexuses*. Cambridge University Press, London.

Compton, R. R. (1962) *Manual of Field Geology*. Wiley, New York.

Conybeane, G. L. P., and Crok, K. A. W. (1968) *Manual of Sedimentary Structures*. Geology and Geophysics Bulletin 102, Bureau of Mineral Resources, Canberra.

De Jong, R. E., and Scholten, R. (Editors) (1973) *Gravity and Tectonics*. Wiley, New York.

Dennis, J. G. (1967) *International Tectonic Dictionary: English Terminology*. Memoir 7, American Association of Petroleum Geologists, Tulsa.

Dennis, J. G. (1972) *Structural Geology*. Ronald, New York.

Dennison, J. M. (1968) *Analysis of Geological Structures*. Norton, New York

De Sitter, L. U. (1964) *Structural Geology* (2nd Edition). McGraw-HIll, New York.

Geikie, J. (1953) *Structural and Field Geology* (10th Edition). Oliver & Boyd, Edinburgh.

Goguel, J. (1962) *Tectonics*. Freeman, San Francisco.

Hedberg, H. D. (1976) *International Stratigraphic Guide*. Wiley, New York.

Higgs, D. V. and Tunnel, G. (1966) *Angular Relations of Lines and Planes* (2nd Edition). Freeman, San Francisco.

Hills, E. S. (1953) *Outlines of Structural Geology* (3rd Edition). methuen, London.

Hills, E. S. (1972) *Elements of Structural Geology* (2nd Edition). Chapman & Hall, London.

Hobbs, B. E., Means, W. D. and Williams, P. F. (1976) *An Outline of Structural Geology*. Wiley, New York.

Hubbert, M. K. (1972) *Structural Geology*. Hafner, New York.

Johnson, A. M. (1970) *Physical Processes in Geology*. Freeman, San Francisco.

Kent, P. E., Satterthwaite, G. E. and Spencer, A. M. (Editors) (1969) *Time and Place in Orogenic Belts*. Special Publications 3, Geological Society of London.

Lahee, F. H. (1961) *Field Geology*, McGraw-Hill, New York.

Leith, C. K. (1923) *Structural Geology* (2nd Edition). Holt, New York.

Lovorsen, A. I. (1960) *Palaeogeologic Maps*. Freeman, San Francisco.

Low, J. W., 1957. *Geologic Field Methods*. Harper and Row, New York.

McClay, K. E. and Price, N. J. (Editors) 1981. *Thrust and Nappe Tectonics*. Special publication 9, Geological Society of London.

Moore, C. A. (1964) *Handbook of Subsurface Geology*. Harper & Row, New York.

Moseley, F. (1981) *Methods in Field Geology*. Freeman, San Francisco.

Moseley, F. (1979) *Advanced Geological Map Interpretation*, Arnold, London.

Newall, G., and Rast, N. (Editors), (1970) *Mechanism of Igneous Intrusion*. Special Issue 2, Geological Journal, Liverpool.

Pettijohn, F. J. and Potter, P. E. (1964). *Atlas and Glossary of Primary Sedimentary Structures.* Springer, Berlin.

Philips, F. C. (1971) *The use of the Stereographic Projection in Structural Geology* (2nd Edition). Arnold, London.

Price, N. J. (1968) *Fault and Joint Development in Brittle and Semibrittle Rock.* Pergamon, Oxford.

Ragan, D. M. (1973) *Structural Geology: an Introduction to Geometrical Techniques* (2nd Edition). Wiley, New York.

Ramberg, H. (1981) *Gravity Deformation and the Earth's Crust (2nd Edition).* Academic Press, London.

Ramsay, J. G. (1967) *Folding and Fracturing of Rocks.* McGraw-Hill, New York.

Robertson, E. C., (Editor) (1972). *The Nature of the Solid Earth.* McGraw-Hill, New York.

Russel, W. L. (195 *Structural Geology for Petroleum Geologists.* McGraw-Hill, New York.

Shrock, R. R. (1948) *Sequence in Layered Rocks.* McGraw-Hill, New York.

Spencer, E. W. (1977) *Introduction to the structure of the Earth* (2nd Edition) McGraw-Hill, New York.

Turner, F. J. and Weiss, L. E. (1963) *Structural Analysis of Metamorphic Tectonics.* McGraw-Hill, New York.

Vistelius, A. B., (1966) *Structural Diagrams.* Pergamon, Oxford.

Voight, B. (editor) (1976) Benchmark Papers in Geology: *Mechanics of Thrust Faults and Décollement,* Downden, Hutchinson & Ross, Stroudsburg.

Weiss, L. E. (1972) *The Minor Structures of Deformed Rocks: a Photographic Atlas.* Springer, Berlin.

Windley, B. F. (1977) *The Evolving Continents.* Wiley, New York.

Whitten, E. H. T. (1966) *Structural Geology of Folded Rocks.* Round McNally, Chicago.

List of Geological Survey Maps

Institute of Geological Sciences (United Kingdom)
(a) *Geological Survey of England and Wales.*

Scale: 1:63,360 or 1:50,000

*18.	Brampton (S)
*23.	Cockermouth (S)
28.	Whitehaven (S)
70.	Leeds (S)
*72.	Beverley (D)
*112	Chesterfield (S & D)
121.	Wrexham (S)
137.	Oswestry (S)
*152.	Shrewsbury (S)
*166.	Church Stretton (S)
180.	Wells (S & D)
*182.	Droitwich (S & D)
226 & 227.	Milford (S)
228.	Haverfordwest (S)
*233.	Monmouth
244 & 245.	Pembroke & Linney Head (S & D)
*251.	Malmesbury (S & D)
264.	Bristol (S & D)
*281.	Frome (S & D)
297.	Wincanton (S & D)
298.	Salisbury (D)
311.	Wellington (D)
327.	Bridport (S)
*342.	Weymouth (D)

Special Sheets

Bristol (S & D)
*Isle of Wight (D)

Scale: 1:25,000

*ST 45.   Cheddar (S & D)
 SH 75.   Capel Curig (S & D)

Special Sheets.

Cross Fell Inlier (S & D)
Central Snowdonia (S & D)

(b) *Geological Survey of Scotland.*

Scale 1:63,360 or 1:50,000

*15.   Sanquhar (S)
*24.   Peebles (S)
*30.   Glasgow (S)
*31.   Airdrie (S)
*32.   Edinburgh (S)
*40.   Kinross (S)
*44.   Mull (S)
*53.   Ben Nevis (S & D)
 71.   Glenelg (S)
 92.   Inverbroom (S with D).

Special Sheets'

*Assynt (S with some D).
 Arran (S)

Note that the 1:50,000 maps are published as E and W half-sheets with the same sheet numbers.

Unites States Geological Survey
Geological Quadraugh Maps

*19.   Athens
 109.   Bedford
 111.   Duffield
*132.   Timpanogos Cave
 141.   Boulter Peak
 162.   Cameron
 172.   Ewing
 188   Coleman Gap

*Maps of particular use for teaching purposes.

Index